THE REAL STORY OF
RISK

GLENN CROSTON

THE REAL STORY OF

RISK

ADVENTURES
IN A HAZARDOUS WORLD

Prometheus Books

59 John Glenn Drive
Amherst, New York 14228–2119

Published 2012 by Prometheus Books

Cover image © 2012 MediaBakery
Cover design by Nicole Sommer-Lecht

Inquiries should be addressed to
Prometheus Books
59 John Glenn Drive
Amherst, New York 14228–2119
VOICE: 716–691–0133
FAX: 716–691–0137
WWW.PROMETHEUSBOOKS.COM

16 15 ~~14~~ 13 12   5 4 3 2 1

Library of Congress Cataloging-in-Publication Data

Croston, Glenn E., 1964–
    The real story of risk : adventures in a hazardous world / by Glenn Croston.
        p. cm.
    Includes bibliographical references and index.
    ISBN 978–1–61614–660–3 (pbk.: alk. paper)
    ISBN 978–1–61614–661–0 (ebook)
    1. Risk—Social aspects. I. Title.

HM1101.C768 2012
338.5—dc23

                                                                2012023645

Printed in the United States of America on acid-free paper

To my friends and family, to my agent, my editor, and my publisher, and to anyone who's ever taken a risk. And to my friend Rich, who lived fully, risking everything.

"Only those who risk going too far can possibly find out how far they can go."

—T. S. Eliot

CONTENTS

THE RISKY BUSINESS OF LIFE

W e live in a risky world, one packed with danger every-where you look and plenty of places you don't. Every day we get out of bed we're up against cancer, earthquakes, financial crashes, terrorism, crime, heart attacks, car accidents, and oil spills. Hazards lurk in our arteries, in the air, on the freeway, or waiting to hurl down on us from space. The dangers are as big as the sun or as small as a virus, as exotic as a shark attack or as mundane as slipping in the bathtub. They're everywhere. All in all, it's enough to make you want to pull up the covers, close your eyes, and hope the risks will all just go away.

But somehow they never do. No matter how much we hope, wish, and pray, the risks are still there. There's no escaping them. We can avoid and reduce the many risks we face through prudent decisions about how we live and work, but no matter what we do we cannot make them go away entirely. Many are completely beyond our control, which makes them feel even riskier.

It's a wonder we manage to deal with it all. Often we don't.

Often we get the risks completely wrong, even when we know better. We do our best to ignore common hazards like heart disease, which kills one in five people in the United States, while we are terri-fied of sharks, which are mainly a hazard in movies. We squirm with discomfort in the passenger seat of a car and lose all fear when we're behind the wheel. A man will go to great lengths to protect himself financially with insurance, investments, and savings, yet cross a busy street through traffic in a subconscious attempt to impress a woman. We shrink in fear from snakes, which affect a handful of people each year in the United States, but somehow in 2006 we were blind to the looming financial crisis until it came crashing down all around us. This story plays out over and over many times in many different ways. We're preoccupied with dangers that almost never affect us while ignoring huge risks that affect millions. How we see the world is frequently and wildly out of whack with the real world all around us.

The risks are easy to see looking back, but we're stubbornly resistant to seeing them at the time. How could we not have seen the financial crisis coming? How blind were we? Millions of risky home loans were pushed through with minimal scrutiny and then cut into pieces, repackaged, and sold as if they had been washed clean. Even financial professionals either missed the dangers or chose not to see them. Was the whole country, the whole world even, collectively out to lunch? How can we do better next time at seeing and responding to other risks that are still out there, like climate change, taking steps to prevent them before they explode in our faces?

THE PAST

The answers are rooted deep in our nature and far back in time. We are marvels of evolution, with our brain and body capable of wonders we scarcely understand, but we are also ill prepared in many ways to deal with the world we face today. The problem is that while we are marvels of evolution, we evolved in a different world. Our human nature is not adapted for the unnatural world we now live in.

Humanity's roots lie in eastern Africa in a long string of ancestors that emerged over the last few million years. Gradually over time, these ancestors became more and more like modern humans, arriving at our present biological state as *Homo sapiens* about 160,000 years ago. We did not get this far without some well-honed tools.

One set of tools, called "affective responses," help us to feel our way around with quick-and-dirty, gut-level reactions based in biology, instinct, and emotion. We are constantly scanning the world around us for anything that feels risky or safe; the former leading us to feel fear, disgust, and outrage, and the latter stirring pleasant feelings like security, trust, and affection. When you get cut off on the highway, are chased by a dog, or receive a terse e-mail from your spouse, your affective responses sound the alarm. Your heart pounds and your palms sweat as you instinctively mobilize for action. The touch of a loved one, a warm meal, or a good story with friends around a campfire can have the opposite effect, infusing us with the feeling that everything will be okay. Responding this way takes little effort—it comes naturally to us.

We also use our intellect to think through risks on occasion, reasoning our way around hazards using statistics, science, and formal logic to look for better solutions than feelings alone can provide. Thinking things through is slow compared to feelings, and it takes more effort. While the solutions provided by thinking can be superior, the time and effort involved can lead to "analysis paralysis" if carried too far. Thinking through risks can also lack emotional impact and fail to jolt us into action.

To understand how we see risks today, we need to take a look back in time. Recorded history stretches back only five thousand to ten thousand years or so, leaving the rest of the story deeper in time to be pieced together by archeologists who sift through the rocks and dirt to uncover our story in fossils and artifacts. Humans evolved over millions of years, leaving behind scattered fossils that tell the story of who we were and how we got here. Fossils reveal not only physical facts about our ancestors' height, age, and sex, but also clues about how they lived and how they died. The story these fossils tell is not a pretty tale of a happy and simpler halcyon age with humans frolicking in nature's plenty. Their fossilized remains instead reveal signs of disease, with bones warped from lack of nutrition. Fossilized bones often show evidence of predators attacking early humans, leaving holes cut into their skulls by fangs. The world for humans back then would have been like one giant food court with you on the menu.

All of these risks shaped the evolution of humans, pressuring the early species to evolve over time to deal with these risks. With the gene pool constantly eroded around the edges by predators and the hunt for food, the pool moved and shrank. By looking at the genetic diversity of humans on earth today, we can tell that, at one point early on, the human population dwindled to only five thousand or so people. That's about the size of a small liberal-arts college, just without the buildings, books, and faculty, and with far more predators. Despite the dominance of humans on earth today, it was by no means a sure thing that we'd be here at all—we just barely missed extinction. Those few who survived this bottleneck and the many other challenges over millions of years passed on their traits to you, me, and the other seven billion people on earth today. Those who failed to deal with these ancient threats disappeared, and their genes went with them, leaving only their broken, scarred, and fossilized bones.

We are the descendants of the winners in this ancient struggle, the ones who fought these ancient risks and won. We are the children of the survivors, walking the earth here today. We have within us, woven throughout our minds and bodies, all the adaptations that helped them survive. We have their upright stance, their ability to find precious food and run long distances to chase game or avoid being eaten. We have their keen vision, the ability to make tools, and the skill to use rich languages for communication. And we have their feelings and analytical responses embedded in our senses, intuitions, and thoughts.

As the descendants of those who fought ancient risks and won, we are ideally suited to survive the perils our ancestors faced. Unfortunately, though, we no longer live in their world. Their world no longer even exists. The adaptations humans evolved with—our bipedal stance, large cerebrum, opposable thumbs, tool-making skills, symbolic language, and complex social groups—gave us the ability not just to survive in the world but also to shape it toward our own ends. And we have been spectacularly good at this task, reshaping the world almost beyond recognition.

When humans arose they were a few scattered groups in a vast sea of wilderness, while today seven billion of us live on our planet and no corner of our world is untouched by us. When humans arose they lived close to the land, getting food provided from the bounty of nature, or starving when nature was less bountiful. Today the majority of humanity lives in cities and never sees where our food originates; instead we buy it in grocery stores and rely on vast supply chains that span the globe. Kids think that eggs come from cartons, not chickens, and that tuna fish grows in cans. Snakes and predators were once a real threat, but in the developed world of urban dwellers, these ancient threats are seldom encountered except in reality television and movies.

THE PRESENT

If you are reading this book, odds are that you live in one of the more technologically and economically developed parts of the world. Your environment, the world you are immersed in every day of your life, is most likely dominated by computers, grocery stores, cars, smartphones, televisions, and buildings. You have water piped in and wastes piped out, most likely. Most people in the developed world have at least some access to medicine. As life has changed, the dangers we face have changed too.

While the hazards early humans faced, like snakes and predators, have faded, new risks that early humans could never have conceived of have taken their place. The world has changed, but our minds and bodies have not, leaving us poorly equipped to handle the risks of the world we live in today. We may know how to reproduce, find calories, and avoid a charging buffalo, but we deal poorly with the risks of public speaking, the fine print in mortgages, and junk food because our nature leads us astray. We failed to stop the financial crisis because deep within us we are still on the lookout for predators and snakes and can't see the modern risks that surround us, staring us right in the face.

Risk is often seen as a matter of statistical tables used by insurance companies and complex models run by hedge funds to stay one jump ahead of the ever-shifting tides of financial markets. But for most of us in our daily lives, the journey into the world of risk is not one of statistics and financial models but an adventure in a distorted world.

Risk is not something we can fully describe based on probabilities, but it is something that we think and feel our way through, moment by moment, every day of our lives: "It does not exist 'out there,' independent of our minds and cultures, waiting to be mea-sured," wrote Paul Slovic and Elke Weber. "Instead, risk is seen as a concept that human beings have invented to help them understand and cope with the dangers and uncertainties of life."[1]

The influences of past dangers are pervasive in our lives today, shaping nearly everything we feel, think, and do. Our biology and instincts protected us from predators and snakes, but now they lead us astray when these perils have dwindled in the developed world (see ch. 1). We do well at jumping out of the way of a charging buffalo, but we do poorly at avoiding the glacial creep of slow-moving threats like climate change that don't feel like an imminent peril (see ch. 2). While we may go to great lengths to avoid threats, sex and love can lure us to take crazy risks we would never consider if in our right mind (ch. 3). Our tendency to learn from immediate experi-ence makes us think rare risks like oil-well blowouts or earthquakes will never happen (see ch. 4); and our ability to adapt to a changing world makes us immune to everyday risks like driving (see ch. 5). Our basic need for control leads athletes and stock traders to create superstitions in risky situations (see ch. 6), while statistics are ignored because we see the world and its dangers through stories (see ch. 7). Our deep-rooted need to be social and belong to the group makes us fear public speaking beyond all reason, and makes us vulnerable to white-collar criminals (see ch. 8). Money, drugs, and food make us run toward dangers rather than away, driven by the never-ending quest for a neurotransmitter high (see ch. 9). Everywhere we look, our lives are twisted by our twisted sense of risk.

THE FUTURE

While we had millions of years of evolution that helped humans to adapt to old threats, we have had only a few decades to adapt to the threats of the modern world. The world is changing so rapidly that our biology can't keep up. Given time, we might eventually evolve to better live in the new world we have created. Evidence suggests that evolution is still underway in humanity with our gene pool on the move, but time is short. The evolution of humanity took place over millions of years, but few of us have that kind of time to wait around for the next stage of risk evolution to unfold.

We don't have to wait for evolution, though. Evolution has given us feelings, instincts, and intuitions that flavor how we see the world and its dangers. Evolution has given us the power of intellect that can provide a useful counterpoint to our gut instinct. We cannot change the tools we are given, but by understanding who we are and how we succeed and fail at dealing with the threats we face, we can choose a better course.

In this book, we'll take a journey through the world and its hazards. Along the way, we'll meet Krishna, the shark-attack survivor; Steve, the air-ambulance pilot; and Anthony, who was there at the birth of the financial crisis. There's Dan, the control-freak comedian; Jennie, the superstitious Olympic gold medalist; and Jill, the cave diver. While we each have our own story to tell and our own risks that we face, we also have much in common because of one simple little incontrovertible fact: despite all our differences, we're all human. For better or worse, we're all facing the world's challenges with the same basic tools that come with being human.

The journey starts far in our past, at the origin of humanity. Fasten your seat belts—we're in for a risky ride.

CHAPTER 1

THE EVOLUTION OF RISK

How Our Ancient Biology Distorts Our Dealings with Sharks, Snakes, and a Changing World

SHARK ATTACK

Sharks are scary creatures, maybe the scariest. The slightest mention of a shark attack snaps your attention to a screen, imagining what it would be like if you were the one facing one of these primeval predators. Few of us have actually been through this, but Wall Street banker Krishna Thompson came face-to-face with a shark and survived to see them in a whole new light.

It was a warm August morning when Krishna swam out from the beach on Grand Bahama Island in the Caribbean. On a vacation with his wife, AveMaria, for their tenth wedding anniversary, he woke early that morning and went out to the beach for a swim while his wife slept in. Hurricane Barry was off the coast, making the water choppy, cold, and murky. Treading water at a depth of about four or five feet, he was the farthest out among the people in the ocean that morning. He was gazing out to sea when he saw a fin moving fast through the water and coming straight toward him. Krishna tried dodging the shark, but it grabbed his left leg, teeth crunching on bone, and started towing him out to sea.[1]

Trying to shake loose, he had no such luck. "I can't believe there's a shark on my leg," he thought, his mind darting to thoughts of his wife and the kids he didn't have yet as he plowed through the water. Later, experts told him that, based on the bite marks on his bone, it was a bull shark that got him that day, one of the few types of aggressive sharks.

"Oh God, get me out of this," he thought as the shark towed him out deeper and deeper and then suddenly pulled him beneath the surface in a swirl into the darkness, shaking his body violently. Tensing, Krishna felt no pain from his leg but worried about not being able to breathe. Knowing he did not have much time, he reached around in the dark to where he knew the shark's mouth must be on his leg and grabbed its jaws to pry them open.

To his surprise, it worked. His leg came free and he was suddenly filled with new energy, happy to be free even if the shark was still right there, staring him in the face. "When you're in the shark's jaws one minute and then you're in front of the shark, you're really happy. I started going crazy, doing combinations, hitting him in the nose, eyes, and mouth. Before you know it, it turned around and swam away."

By this time, the ocean around Krishna was dark red with his blood. He swam back to shore, doing the breast stroke. When he got back to shallow water, he hopped on his good leg toward the beach and onto the sand. He tried to scream but couldn't at first, his body failing to respond due to the loss of blood. When he managed to scream for help, people came, but he remembered little after that for some time. Later he learned that the doctors had a hard time stabilizing him because he'd lost so much blood. They worked on him for hours, from eleven o'clock in the morning until six o'clock at night, his heart stopping more than once. By the time he woke up, he'd been moved to a hospital in Miami, Florida.

Later, Krishna was fitted with a prosthetic leg using the C-Leg, with a microprocessor prosthetic knee made by Otto Bock HeathCare (the *C* stands for *computer*). Though a realistic-looking prosthesis, it hurt to use at first, until he got used to it. In addition to continuing his work as a banker, these days Krishna also talks to people about prosthetics, blood donation, and shark conservation. Yes, shark conservation. Recently he was part of a group of shark-attack survivors testifying to the US Senate about shark finning, the practice of catching sharks just for their fins, which has decimated shark populations.

"The reason why I work for sharks is that it's not about the shark or me—it's bigger than that," said Krishna. "It's an issue that needs to be addressed and needs to be addressed now. Sharks are being depleted, they're slow growing, mature late, and have few offspring. At this rate, they'll be extinct soon, and we need sharks in our water, for our oceans and our world. Sharks have been around for hundreds of millions of years. If sharks die, the oceans die; and if the oceans die, then we're next," he concluded. It's more than a little ironic that the greatest risk we face from sharks isn't that they will attack us but that they won't be there at all.

When you take a look at a shark's mouth, it's not hard to see why they're so scary. The jaws of a great white hold hundreds of jagged teeth, rows and rows of them all the way around its mouth. In our mind, sharks are merciless monsters, unstoppable eating machines constantly prowling the depths for those unlucky enough to wander into their domain. There's just one problem with this. Despite the ordeal that Krishna Thompson and others have gone through, shark attacks are rare and fatalities are even rarer. Knowing this does not seem to make sharks any less frightening, though.

Across the United States, sharks killed twenty-five people from 1959 to 2008, while lightning (a rare risk itself) killed 1,930

Americans over the same time period.[2] Pet dogs kill more Americans than sharks. Worldwide, shark attacks claim a handful of lives every year. For every death by a shark, heart disease claims over six hundred thousand lives. And yet a shark attack always makes the news, no matter how little it has to do with our lives. We can live in Kansas, never setting foot in the ocean in our lives, and still shiver at the thought of a shark attack. We're programmed to be interested, to see the risk of a bloody demise by tooth and claw as a grave risk, no matter what the numbers tell us.

The stories of shark attacks go back deep in history. The Greek historian Herodotus wrote of shipwrecked Phoenician sailors getting eaten by "monsters" in the Aegean around 490 BCE.[3] There are sharks in cave paintings and on totems of Pacific Coast natives of North America. Native Hawaiians have stories of sharks in their culture, including the King of Sharks who took human form and married a beautiful woman, only to have their child driven from the island to take shark form again when the fishermen feared the boy was chasing their fish away.[4]

And then there's *Jaws*, which came out in theaters in the summer of 1975. A whole generation stayed out of the water and abandoned the beaches in droves after watching Bruce the shark go berserk in theaters. *Jaws* may have amplified the fear of sharks many feel, but it did not invent it. It exploited a fear that was already there, the fear of the unknown lurking in the depths, the fear of being eaten alive by a ruthless predator.

"Our innate fear of sharks comes from our inability to protect ourselves in the marine environment," Ralph Collier, founder of the Shark Research Institute, said in an interview.[5] "Even in clear water, we can only see so far down in the water, while something from below can see us against the backlit sky. The fact that there's an animal that can eat us alive and there's nothing we can do about it is always in the back of our mind. In water we're out of our element, out of our realm. We're totally vulnerable to anything. You have no control."

Collier has spent most of his life studying sharks. As the founder of the Shark Research Committee in Southern California, he's been studying sharks for fifty years both in and out of the water. In the early days of his research, little was known about sharks, which was one reason he started his work in the field.

"People said that people and divers were attacked because they looked like seals, but that's not true," said Collier. "Sharks have very good vision, with the same rods and cones that we [have], and the same visual acuity and color detection. Because they have acute vision, they don't mistake a surfboard for a seal and could not possibly mistake a diver for a seal. From the types of bites to surfboards and divers, probably most are from investigation. White sharks bite a lot of things. Crab-pot buoys are yellow and blue and don't resemble anything in the ocean—sharks bite them all the time." Collier estimates that 85 percent of shark attacks are the result of sharks' investigation or curiosity, not predation.

There are stories of boats getting bumped by sharks, or even of sharks knocking boats over, like malicious demons of the depths. But these stories are mostly about sharks just pushing people back, says Collier. "The boat gets in the way, and suddenly the boat is in his area. When [the boat pulls] away, the shark goes back and continues to feed." It's as if the shark feels that you're in his personal space and wants you to back off. He probably feels how you feel when somebody crowds you in the elevator or sits too close on a park bench.

The world has changed a lot since Collier started his research. For one thing, the number of sharks in the oceans has plummeted. In truth, we're more of a threat to sharks than they are to us. "Some populations are off by 90 percent," said Collier. "At least a third of sixty species of pelagic sharks are endangered. We are facing a critical problem. The finning of sharks on the US East Coast has taken so many sharks of so many species that the cownose ray population has exploded. The only predator of the rays is sharks. Without sharks, if you could get all of the rays to come to the surface, you could walk across [the Chesapeake Bay]. It's decimating the shellfish industry, setting up an imbalance in the ecosystem, and eventually it would end up with the whole system collapsing." Ultimately by killing sharks, we're a threat to ourselves.

Sharks might have the last laugh, though. Rather than getting their revenge by eating people, the real risk comes when we eat them. "Like most people, I like fish," said Collier. "And I used to have a shark steak now and then until they studied heavy-metal poisoning and found high levels of methyl mercury in shark meat, and even higher levels in shark fins." Methyl mercury is a potent poison derived from mercury released into the environment by people, which then concentrates in sharks and other fishes. In recent years, huge ships have hauled in thousands of sharks at a time, and workers have removed their fins to sell them and then dumped the sharks back into the ocean to sink to the bottom and drown. As more shark fins are taken, more people eat them and are exposed to the mercury they contain. "And now we're finding that because of all the shark fins on the market, their cost has gone down, availability has gone up, and now you can go into most restaurants in Japan or elsewhere in Asia and . . . buy soup. People who eat shark-fin soup every day will never get rid of it [methyl mercury in their body]. We're seeing neurological disorders and birth defects, all traced to shark-fin soup."

The sharks lose; the humans lose—everybody loses. It's kind of your classic lose-lose situation, I guess.

DIGGING IN THE DIRT

How we see risk is the result of millions of years of evolution that shaped our species, making us who we are today. Evolution never sleeps, tirelessly working to shape and change the millions of species on earth, including humans. Relentless evolutionary pressure shaped generation after generation of vertebrates, mammals, and primates long before our ancestors were human or anything like it. To know how our environment shaped our ancestors and ultimately made us who we are, we have to understand what that environment was and what kind of risks this environment held. And to do that, we have to get down and dirty.

Much of the evidence we have about our past comes from digging in the dirt. Archeologists and anthropologists make a living out of getting dirty, carefully sifting through tons of dirt, sand, and dust for rare fragments of fossilized bone, tools, trash, and any other sign of early humanity. The dirt of the valley in the East African Rift has been a particularly rich trove of human fossils, and it is also revealing clues about the environment our ancestors lived in at the time, including coexisting plants and animals, as well as the climate. As you move forward in time through the layers of the fossil record that bring us closer to the present, more clues to humanity's origins appear, such as tools, shelters, clothing, art, and burial sites.

Fossils from a long string of human ancestors are found in the same general area of eastern Africa, originating from as far back as about six million years ago on up until the time modern humans, or *Homo sapiens* (that's us), appeared on the scene about 160,000 years ago. Altogether, the remains of about six thousand individuals, usually only fragments of bones, have been found spanning this time. At the earlier end of that time period, hominid species like *Sahelanthropus tchadensis* still looked quite apelike overall, and their brain was smaller than a chimpanzee's. But based on how their skull sat on top of their spine, we can surmise that they walked upright. As time progressed, a variety of human species came and went in the budding tree of humanity, often with a few different humanlike species living at the same time and even in the same place.

The popular story of early humanity often portrays humans as wily and aggressive hunters on the primeval scene, wielding spears against mammoths in the snow. And while our primate ancestors may have been smarter than many of the other creatures around them at the time, they could not have been the fiercest or fastest animal. They lived in a world with giant hyenas that weighed 440 pounds and saber-toothed cats with enormous slashing teeth like knives. The humans, meanwhile, had no fangs or claws to speak of and were thin skinned with little in the way of defenses. Dr. Robert Sussman, professor of anthropology at Washington University in St. Louis, Missouri, and his coauthor Donna Hart, professor of anthropology at the University of Missouri in St. Louis, have come to see early humans as more often the victims of predators, as described in their book, *Man the Hunted: Primate, Predators, and Human Evolution.*[6] All in all, early humans were a tasty little meal for many predators.

"The story of man the hunter was really in the late Pleistocene, about forty to sixty thousand years ago," said Sussman in an interview.[7] By this time, modern humans were on the scene and were systematically hunting their prey, including large animals like mammoths, with the aid of tools, like spears. They were also better able to defend themselves, although even then predators remained a great challenge. The challenge was even greater for our more ancient ancestors, who would have been practically defenseless, lacking anything but the most primitive of tools. But this is only the most recent slice of human prehistory.

"Around two to five million years ago, when humans were becoming more human than apes, they were about four and a half feet tall, and for them the world must have been a very scary place," said Sussman. "The predators at the time were five times as large, and ten times more abundant, than now. Even today, for medium to large primates, the main danger other than humans is predation, and early humans were largely unprotected."

Early humans did not have much in the way of tools until about two million years ago, which came in the form of pebble tools, primitive hand tools consisting of a rock with the edges chipped off. Imagine yourself with a sharp rock in your hand, facing a lunging leopard far larger than yourself. Not a pretty picture. These tools fit in your hand and might have been handy for small prey or for cutting things, but they were worthless for protection from large predators lunging at you or for hunting larger animals.

The fossilized remains of humans often show evidence of predator attacks from the likes of lions, hyenas, and even giant crocodiles, with as many as 4–6 percent of fossils showing signs of predation, Dr. Sussman explained. The bones of these early humans reveal bite marks, scrapes, or even puncture wounds left by talons or claws. Large predatory cats alive from one to two million years ago assembled piles of bones from their prey in South African caves, including primates and early hominids. Besides cats, a host of other predators, like hyenas, birds of prey, crocodiles, and even giant snakes preyed on humans. "Paleontological evidence supports the conclusion that both hominids and other primates, such as baboons, were frequent meals for ancient predators," writes Sussman, describing case after case of fossil remains that showed signs of predation, like holes in skulls where animals' fang-shaped teeth must have gripped their hominid prey.[8]

One of the hominids in Africa dating back to around 2.7 to 4 million years ago was *Australopithecus afarensis*, a species that lived at the edge of forests and grasslands in social groups, as their fossils indicate. Their teeth suggest they had a vegetarian diet, with flat molars for grinding rather than the sharp, pointed teeth needed for eating meat. One of the most famous human fossils, "Lucy," was a member of this species that lived 3.2 million years ago. *A. afarensis* fossils also indicate that, as a species living on the edge of the forest, they often fell prey to predators, as is the case with modern edge-dwelling primates. It's unlikely that unimposing hominids like Lucy were out hunting meat, and it is far more likely that they were hunted.

The threat of predators isn't completely relegated to the remote past. One of the biggest risks for a broad variety of primates on earth today, other than humans, is still predators. They're a risk not just for smaller primates but larger ones, such as chimpanzees, as well. Chimps are not necessarily cute and cuddly; they are large, powerful animals five to seven times stronger than humans, with a sharp set of teeth. Field observers have reported chimps fighting off or killing predators when attacked, but this does not always prevent wild chimps from being preyed upon by leopards, lions, and other predators to the tune of about 6 percent of their population each year. Baboons and even mountain gorillas also fall prey to feline predators like leopards, so size alone would not have protected early humans from predators. Humans were big enough to provide a good meal but not large enough to put up a big fight.

In some parts of the world, predators are still a threat to humans. Although their original geographic range has fallen by an estimated 93 percent because of hunting and habitat loss, tigers still kill people in areas like the Sundarbans delta of the Ganges River in India and Bangladesh, with hundreds of people falling prey to tigers over the course of a decade.[9] Reports of joggers and hikers meeting mountain lions have grown as suburbia has expanded into wilder areas in the western United States. So predators do attack sometimes, but they are not a major threat for most of us.

Our fear of predators is rooted more deeply than warranted by statistics or experience concerning this threat. We retain a deep-rooted fear of predators that media and movies exploit to the fullest of their abilities. But Hollywood did not invent this fear—it was already there long before the movies were around to capitalize on it. While today we live in a world of concrete and cars, the fear of slashing teeth and claws is still deeply embedded in us. Images of predators attacking send instinctive chills down our spines. When you are walking through a dark wooded area at night, or even walking down a darkened city street, feeling on edge and surrounded by shadows, it's as if the echo of a past age is still with us, looking for predators that might leap out at any moment. Predators touch a sensitive evolutionary nerve, one that goes back millions of years for us. When we shiver at the thought of a man-eater, repulsed by the concept of being eaten alive, the feeling runs deep.

A wide variety of animals have an instinctive fear of predators, or at least a predisposition to learn to fear specific predators, even if the fear is not automatic. If you take a lab mouse that has never seen a cat and give it a whiff of cat urine, it instinctively reacts, freezing up to stay unseen. Humans everywhere seem to respond in the same way to images of predators in action; the thought of being eaten alive requires little analysis to stir deep fear and anxiety. "After all, we evolved for millions of years being hunted and eaten by other animals, but we have only had to fear automobile accidents for 100 years (just a few generations)," writes Sussman. "Tigers, bears, and wolves touch off much deeper neural pathways than Toyotas, Fords, or Volkswagens."[10]

SOCIAL SECURITY

Predation exerts a strong selective pressure on populations, including early humans. Natural selection would cause those with better defenses against predators to survive and reproduce more frequently. Many parts of our human nature may have been influenced by the need to deal with predators, eons before humans started hunting. Standing upright on two legs, making tools, and getting progressively more intelligent may be, at least in part, the result of predation. And predators may have been a key evolutionary force that drove humans to develop as social animals, eventually building cities, societies, nations, and cultures with billions of people.

What drove our social development is that there's safety in numbers, which is one big reason why so many animals live in groups, like a gaggle of geese, a school of fish, and a herd of asses. Furthermore, humans are not the only social primates. Baboons, chimps, gorillas, and every other large primate species live in social groups of varying size and complexity, and the members of these groups work together to protect themselves from predators like leopards. The complexity of human society may be greater than that of these primates, but its roots are probably the same.

The impact on early humans, who were very prone to being eaten, would have been dramatic. They had to use their wits and ability to cooperate in order to survive, and those with more wit and a greater ability to cooperate were probably more successful and came to dominate the gene pool.

In a group, you have more eyes to help scout and alert each other to a predator, and even fight them off, if necessary. Predators work hard to single out an individual from the pack, knowing that on its own, an object of prey is far more vulnerable. This works in favor of groups of gazelles or zebras protecting themselves from lions, and also for primates today.

When researchers looked at how different primates live in social groups, they found that those living in larger social groups were exposed to more predators. "When we look at the number of individuals in a primate group and the corresponding predation rate, it suggests that predation may have played a key role in the evolution of basic primate attributes, such as group living," writes Sussman.[11]

The fear of predators may even affect how we see open spaces in art and in the great outdoors.[12] Landscape architects, park managers, and artists know that we love sweeping open areas and broad vistas, with running meadows stretching away to distant mountains. We love an open landscape where we can climb a hill or a mountain and peer out over the countryside below. Forests, on the other hand, are often the focus of darkness and evil in stories. You cannot see far ahead of you in the forest, and danger lurks behind every tree. Think of the hobbit Bilbo Baggins and his encounter with ginormous spiders in a dark, creepy forest. Or think of Harry Potter and his own encounter with giant spiders in his own creepy forest. All in all, forests are bad news, in stories at least. All of this might stem from an adaptation to look out for predators in the landscape. When you look at a sweeping, panoramic view and feel a stirring of pleasure, perhaps it is the pleasure of feeling secure, looking far in all directions and seeing no predators anywhere in sight.

WHY DID IT HAVE TO BE SNAKES?

Snakes are rarely a threat in North America or Europe, causing fewer than twelve deaths a year in the United States,[13] but snakes are still a significant hazard for people in many parts of the world. Tens of thousands of people die each year globally as a result of venomous snake bites, particularly in the tropics of Asia and Africa. Snakes were also a threat to our primate ancestors for many millions of years, and their imprint is still inside us, it seems, giving us a deeply engrained predilection to fear snakes regardless of whether they are a real threat today.

We like to think that we are above instinct, but when faced with snakes, the fear is so common that it's hard to explain without a deep-rooted response that stems from biology. A Gallup poll of adult Americans found 56 percent listing snakes as one of their greatest fears, right up there with public speaking.[14] It doesn't matter if you're from Boston or Alaska and the only venomous snake you'll ever see lives in a zoo. The fact that this fear is so abundant for something we seldom encounter suggests that, deep down in our wetware, we are wired to fear snakes.

Humans are complex animals; it can be difficult to piece together whether a behavior in humans is truly instinctive, learned, or somewhere in between. You can't raise a human baby in the absence of all snakes, snake pictures, snake toys, snake stories, or anything snake-like, and then stick a snake in the child's face to see if it cries. We might be born with a fear of snakes, we might have a predisposition to develop that fear, or we might have an innate fear that some people override. But one way or another, snakes give us the willies.

The threats from snakes include bites from venomous snakes and the capability of extremely large snakes to actually become man-eaters, no less fearsome. Dr. Sussman describes in his book an experience he had while studying lemurs in the jungle of Madagascar, where he encountered one of the giant snakes. "Stepping on a large branch, I felt it begin to move. Letting out a scream, I nearly jumped out of my shoes. I had stepped not on a branch, obviously, but on an extremely large Malagasy Boa."[15]

Personally, I'm not all that afraid of snakes, but I'm also not immune from their impact. If I'm out on a trail and see a snake, I'll stop and take a look or take a picture rather than run away, but I'll still keep a respectable distance. Rattlesnakes are not uncommon in the part of North America where I live, San Diego, California; and several hundred bites occur every year according to public State of California figures. But bites are seldom lethal if you get medical care quickly, with only one to two deaths per year. The facts about the risk from snakes somehow don't seem to mean much when you run into one, though. Once while jogging on a trail in a local canyon, I came to a creek running across the trail. The water was not wide, so I decided to jump across. As I jumped over the water, I heard a rattle midstride and looked down to see a rattlesnake half-submerged in the water beneath me, its head and tail sticking out. It never occurred to me before that moment that they could swim. Needless to say, I jumped extra far over that creek and ran quite a bit faster for the rest of the run that day.

Stories of snakes pop up in cultures around the globe as symbols of fear and evil in everything from the Garden of Eden in the Bible to the *Indiana Jones* movies. Snakes may not be a great risk to North Americans or Europeans today, but they remain a significant threat in many parts of the world, and the risk from snakes was even greater in the past. In fact, snakes may have been a great enough risk to drive primate evolution and make us who we are.

Monkeys share our fear of snakes, but not from birth. Monkeys raised in the lab without ever seeing a snake don't appear to be instinctively afraid if shown one later, but a wild monkey shown a snake will shriek and cower. If the lab monkey sees the wild monkey respond with fear toward snakes, it too will be afraid the next time it is shown a snake picture, a toy snake, or an actual snake.[16] It learns to be afraid.

Lab-raised monkeys cannot be taught to fear just anything this way, however. If a lab-raised monkey sees a wild monkey showing fear toward flowers, it won't freak out when the next floral arrangement arrives.[17] Even if the fear is not entirely predetermined, the predisposition for the fear of snakes seems to be waiting to be triggered in monkeys and perhaps in us. There's something specific about snakes that monkeys and humans have evolved to fear—perhaps that curved, slender shape of theirs. Our minds are wired to find that snake shape as we scan the world around us, and when do we are pre-wired to fear it more than other things we see. We tend to stay on the lookout for snakes and easily develop a fear if the situation arises.

Dr. Arne Ohman of the Karolinska Institute in Sweden has spent years studying our fear of snakes and spiders and was afraid of snakes himself as a youth. Rather than exposing people to venomous snakes, Ohman and his colleagues showed people pictures of snakes or spiders hidden in a three-by-three grid with pictures of mushrooms and flowers, then measured how quickly subjects could find a picture of a snake or one of the other images.[18]

The theory is that if we have an innate ability to detect threats like snakes, then we might be able to find them faster than the other objects' pictures in a test like this. Snakes do seem to have a way of grabbing our gaze, even causing us to mistake something like a curved stick lying on the ground for a snake. When Ohman did his experiment, he found that people could find pictures of snakes or spiders more rapidly than they could find pictures of mushrooms or flowers when the pictures were arranged together in a matrix. The response time to find the picture was only a second or so, but consistently in a large number of tests, the same thing kept happening, with the snakes winning out. Taking the experiment a step further, Ohman tested people to find individuals with a fear of snakes or spiders. Repeating the experiment, he found that people specifically afraid of snakes were particularly rapid in finding snakes in the matrix, but did not have the same ability to find spiders. Our fears can change how we see the world, making us particularly good at finding the things we fear most, at least when it comes to snakes.

When we feel fear, our skin immediately and automatically and produces a thin layer of sweat as our nervous system kicks in with the fight-or-flight response. Our instinctive responses, like this one, are more honest than the words we say, which is why the polygraph test reveals lies by measuring this sweat response. Your skin conductance shows your fear even if you do not know how or why you are afraid and aren't even aware of it. Ohman measured the skin conductivity of people who were shown images of snakes for as little as thirty milliseconds, so briefly that those in the study were not even aware they had seen a picture of a snake.[19] Their skin knew it, though, getting the signal from deep within the brain, deeper than higher thought centers involved in thoughts that we are consciously aware of. They could not verbalize that they had seen a snake, but those who were afraid showed it in their skin.

When you see a threat like a snake, you don't have a lot of time to think through your options. You need to act immediately, and an ancient part of our brain, called the amygdala, helps us to do this. The visual information about a snake goes straight to the limbic system in the brain, an ancient set of wiring including the amygdala that plays a key role in fear responses in vertebrates, including humans. To ensure a speedy response, processing by the amygdala does not require first going through higher thought areas in the cerebrum, the part of the brain we normally associate with higher thought. Instead, the amygdala jumps right into action without thinking through all the options. The amygdala may not be very smart or a great learner, but it is reliable and fast, responding to other ancient threats like angry faces in the same way it responds to snakes. Act first and think later is how this ancient system works.

The amygdala includes two small, almond-shaped regions deep in the brain that light up in brain scans when we are exposed to things we fear. Animals without an amygdala appear to lose all fear, which is one reason scientists have a pretty good feeling about what the amygdala does. Monkeys with their amygdalae removed lose all fear of snakes and spiders, playing with the creepy crawlers instead of cowering or shrieking. This experiment is hard to reproduce in humans; however, a woman known to the world as "patient SM" was born with a rare condition with a very selective loss of her amygdala, providing a glimpse of what life without fear might look like.

SM's curious life has been the subject of many studies looking at the role of the amygdala in human fear.[20] She has trouble recognizing the emotion of fear in pictures of faces and seems to have no fear herself, despite feeling and recognizing the full range of other emotions. In a haunted house, she startled one of the actors playing a monster rather than the other way around. After being attacked in a park by a mugger, she continued through the same park the next day, same as always, without giving it a second thought, unaware that this might be unusual. And when SM was shown snakes and spiders in a pet store, she gravitated toward them rather than shying away. There's something very basic, it seems, in our fear of snakes.

"We have a natural tendency toward fear of snakes but we also have huge brains that can override that tendency," said Dr. Lynne Isbell of the University of California at Davis, located in the Central Valley of California, near Sacramento.[21] Though known as "Aggies" in reference to the school's agricultural roots, those at UCD do a lot more these days than raising pigs and tipping cows. Dr. Isbell does fieldwork in

Kenya studying the behavioral ecology of monkeys. The monkeys are quite skilled at seeing snakes, and when they do, they let everyone else know, including Dr. Isbell. "I count on the monkeys—they see snakes far faster than I can and they give alarm calls to them, so I don't worry about snakes. I worry about Cape buffalo and elephants. The monkeys do not give alarms for these animals, and they will charge without apparent provocation," Isbell said in an interview.

In her book, *The Fruit, the Tree, and the Serpent: Why We See So Well*, Dr. Isbell argues that snakes helped to shape the evolution of mammals and primates that, in the evolutionary tree, eventually led to early humans. Snakes might be feared, but the selective pressure they created through evolution might be the reason we are here today.[22]

"There's an incredibly long history of our lineage dealing with snakes," said Dr. Isbell. "Venomous snakes don't eat most primates, but they're still deadly. Even though society has removed the threat of venomous snakes from our lives here in the US, we're still dealing with the result of sixty million years of living with venomous snakes. It's been such a strong selective pressure to deal with snakes that snakes have caused primates to be primates."

Part of what makes us human is our eyes and the mental processing power used to determine what we see. One way snakes shaped us may have been our unusually acute vision. We have large, forward-facing eyes on our head, great for binocular vision. Our vision is unusually acute in terms of colors, as well as depth, to detect snakes against a varied background. And our vision is wired to our amygdala in our brain, the fear center, which seems to be important for both detecting snakes and becoming afraid of them.

Some who study human evolution have argued that the acute vision we have was the result of searching for food or clutching for a hold while primate ancestors were climbing in the trees, but these arguments have holes. Early primates probably ate a variety of food, not just small food like insects that required fine-tuned vision and gripping dexterity.

The arms race with snakes may also have helped primates and, later on, humans with the increasing size of their mental-processing capacity. We developed extensive visual-processing capacity that has served us well. We can read and write music, read the fine print on mortgage documents (sometimes), and appreciate a painting all thanks to our vision and visual-processing capacity, as well as our higher mental functions.

Snakes may have even helped humans to develop language. Gesturing is a simple form of communication and might have been a precursor to language. And one of the simplest gestures is pointing a finger. It turns out that there is more than one way to point your finger, not counting the one you're thinking of right now. One type of pointing is called *declarative pointing*. Whenever we point our index finger at something to share this information and attention with others, we are performing declarative pointing: telling others to look and see what we see. Infants as young as twelve to eighteen months perform declarative pointing to share with adults, using this simple, but effective, form of communication.[23]

As Dr. Isbell explained in her interview,

> We have the ability to point to objects that we want to share interest in with other people. Chimps can point, and dogs can point, but to things that they want, not to share interest. Why did we evolve that ability to point to things to share interest? Children with autism don't point in this way, and they also have difficulty with communication and social relationships. There's a connection between the rate of language acquisition and how early kids start pointing declaratively. Why did we start pointing declaratively? Probably to point to dangerous objects in [our] environment. Snakes would be useful to point at. When we point, our eyes get

drawn to the tip of our finger. It brings our eyes closer to the target. This is especially useful under conditions where the target is obscured within the clutter of vegetation, as snakes so often are. ... So I suggest that snakes have led indirectly to the evolution of language in humans.[24]

While a tendency to learn to fear snakes may be part of who we are, our relationship with snakes can be complex. Some people love the critters, even keeping them as pets. Sometimes snakes attack humans, but sometimes humans and snakes compete for prey, and sometimes the humans turn the tables on the snakes and eat them. Dr. Harry Greene has studied the relationship between snakes and humans in tribes of the Agta people in the Philippines, a hunter-gatherer tribe that, for the most part, no longer exists (at least not living on the land with the same customs and environment they once had). Looking back on records reveals a complex relationship between the forest dwellers in the area and the reticulated pythons with whom they shared the forest. These snakes can be giants, growing as long as five to seven meters in length. When the Agta were interviewed, over one-quarter of adult men in the tribe had been attacked by these snakes at some point in their lives. But oftentimes the humans eat the snake instead, getting as much as forty kilograms of meat from killing one big snake. A more complex relationship like this, in which humans can be both prey and predators of snakes, might explain a more complex range of responses with snakes than simply fear alone.[25]

In her book, Dr. Isbell suggests

that the fruit, the tree, and the serpent have all contributed to make us who we are today. There are other contributors, of course, but without the fruit, the tree, and the serpent at the beginning, we would not have had the foundation of excellent vision, a sense that enables us to develop written languages, draw up blueprints and build complex structures, appreciate the role of color in creating art in all its forms, recognize the difference between good and evil in another person's face, or detect a dangerous snake resting camouflaged in the grass. Thank goodness for snakes![26]

Snakes probably can't take all the credit for our current state. Neurobiologist Mark Changizi argues for a variety of evolutionary influences on our vision and who we are today. He argues that the evolution of primates in forests, for example, is what drives us to have eyes facing forward for binocular vision, providing what he calls "x-ray vision," the ability to see through dense leaf clutter.[27] It seems likely that the evolution of flowering plants also played a role in the evolution of primates. But all in all we did not evolve to build a global technological society. Humans evolved in order to survive the world they existed in at the time. Out of the evolutionary crucible, while escaping snakes and other predators, we emerged with the tools that have allowed us to create the new world we live in today.

OUT OF AFRICA

The past may be a useful indicator about our future, if we look back to Africa once again. The area of Africa where many human fossils are found, the Great Rift Valley, is hot and dry today, but evidence from the past suggests that it may have been quite different during the course of human evolution. Some scientists have suggested that humans evolved in a world that was becoming drier and grassier, and that standing on two feet provided humans with an advantage in how they moved on the ground and how well they could see over the grass when looking for predators. Another view is that the evolutionary forces shaping humanity were not a specific environment of trees or grasses or the difference between hot and cold climates. Instead, this view holds, the driving force shaping humans may have been change itself.

The Great Rift Valley is a creation of plate tectonics, with the valley forming where eastern Africa is being pulled in two, leaving a gash across the continent. A geologically active region, deep lakes span the landscape, going back millions of years. Lake Malawi, also known as Lake Nyasa, is the southernmost lake in the Great Rift Valley and the third-largest lake in Africa, bordering on Mozambique, Tanzania, and Malawi. The lake's waters are populated by a rich variety of cichlid fishes, and its depths go down to 2,300 feet; however, this has not always been the case. Researchers studying Lake Malawi have found that its depth has varied dramatically in the past. Dr. Andy Cohen of the University of Arizona in Tucson has studied Lake Malawi and other lakes in the area, drilling into the mud in the bottom of the lake to explore its history. The layers of mud on the bottom of a lake are not just goopy messes—that goop is packed with information. When researchers very carefully pull up the mud, they can sort through the layers to examine microscopic remnants of pollen and fossils of marine creatures that were alive at the time. The layers are laid down, one on top of the other, giving a history stretching back over hundreds of thousands of years, including the prevailing weather over time. In dry times, less pollen would blow into the lake and get deposited. As the lake dried and became saltier, different plants and animals would inhabit the lake. "Lake Malawi, one of the deepest lakes in the world, acts as a rain gauge," said Dr. Cohen, and this rain gauge has provided him with interesting findings.[28]

While the long-term trend in Africa over the last several million years has been toward slow drying, the trend has not always been in one direction. The Lake Malawi region today is forested and receives abundant rainfall, but it was not always so. From 135,000 to 90,000 years ago, the water level in the lake dropped over two thousand feet, leaving a much shallower and murkier lake. The pollen and signs of vegetation in the area dwindled as a megadrought, worse than any seen in modern times, gripped the region before finally relenting around 70,000 years ago as rainfall increased again and the lake filled back to its present level.

The changes may have been quite rapid, at least on a geologic timescale, with dramatic changes in Lake Malawi occurring over the course of a few centuries. This might not sound fast, but the impact on humans at the time may have been great. "The changes in precipitation in tropical Africa would have had profound impacts on human populations since the food resources people rely on as well as their water supplies all depend on rainfall," said Dr. Cohen. Fossils of humans in the area are few through the period of drought and then rebound afterward. "Foraging for food and obtaining water that isn't too salty to drink under these conditions would have been challenging. Of course people do eke out a living under these types of conditions in Africa today. But the point is that population densities under these conditions are almost always pretty low."

These are not the first changes in the climate of the region. Looking deeper in the lake sediments and at the ocean-floor sediments in the area, Dr. Cohen and other researchers have pieced together a picture of climate change extending even farther back in time, covering the last few million years that were crucial in the course of human evolution. "Some people for example have speculated that the key adaptation of *Homo erectus* may have been an ability to perform sustained running bipedally, which would have been advantageous in the increasingly grassland habitats of the early Pleistocene," said Dr. Cohen.

Dr. Matt Grove from the University of Liverpool has found that periods of rapid climate change over the last few million years coincided with periods of rapid human evolution not just in our bodies but also in the development of tools. One period of climate change about 2.7 million years ago was followed rapidly (relatively) by the appearance of the first stone tools 2.6 million years ago. *Homo erectus* produced only very simple stone tools, a so-called hand-ax, and continued to use this simple tool for almost a million years with little change. *Homo erectus* lasted through this period while other species of early humans did not, perhaps due to its adaptability, living in a broad range and over a period of 1.5 million years. To put that into perspective, our own species has been around only for about 160,000 years, a mere blink of the earth's eye.[29]

Looking at the course of human evolution and comparing this with the record of a changing climate in the area, a surprising chain of connections starts to emerge. For long periods, humans would reach a state of more or less stasis in their biology and culture, and the climate in which they lived would be stable as well. When a shift in climate occurred, or the climate became highly variable as uncovered by sediments in lakes and the ocean, new species of early humans and new innovations in tools would emerge and spread.

This pattern repeated itself more than once. While the discussion was once whether humans evolved in grassland or forest, the true driving force in human evolution was probably change itself. Avoiding snakes and predators may have played key roles in human evolution, but climate change may have been the factor that ultimately put us over the top. The rapidly changing world early humans lived in forced humans to develop the versatility that allowed us to survive a changing world and dominate our changing world today.

It's ironic that climate change shaped us and that we, in turn, shaped our world, creating a new era of climate change. How we deal with this new challenge of modern climate change is another story that is still playing out.

CHAPTER 2

THE SEA OF DENIAL

Why Rising Seas and Heart Disease Are Hard to Stop

SLOW BUT STEADY LOSES THE RACE

Field biologists may not be those whom most people picture when they think of risk takers, but spending so much time in the field, these biologists often find themselves confronting dangerous animals at ground level, armed with little more than their wits and a notebook.

As an evolutionary biologist, Lynne Isbell spends considerable time in the field for research. Scientific research isn't known as an extreme sport, but in the course of her studies on monkeys, Dr. Isbell has had some close calls. For instance, while she was walking in a dusty savannah in Kenya one day, she heard and felt a Cape buffalo charging toward her. Now, although our first response to danger is often to freeze, the second response a few milliseconds later is to fight or to run. Running was a much better option for Dr. Isbell in this case. "I think we have a natural ability to detect and avoid anything that comes at us fast," she said in an interview.[1] The Cape buffalo can stand six feet tall at the shoulder, weigh a thousand pounds, and sport large, curved horns several feet across with sharp points on the ends. They're known for their unpredictable behavior and are sometimes called "the widow-makers" of Africa, killing a hundred people or so a year. When the Cape buffalo charged at Dr. Isbell, she froze for a few milliseconds and then ran. "The fear didn't appear until after the buffalo turned away and I had a chance to think about how close I had come to dying."

Our senses, muscles, and brain are tuned by evolution for fast-moving threats like those our ancestors dealt with. If a buffalo charges at us, we have no trouble detecting the threat and will generally do our best to get out of its way. Our eyes and our reflexes help us to leap aside, followed by the brain's role in planning the next move. Fear helps our lungs and legs to move like our life depended on it as adrenaline surges and a burst of energy is released, the classic fight-or-flight reaction. These days, most people in the developed world don't meet a panther or a leopard outside of a zoo, but if we ever do, we're ready. Far down in our mind, we are still waiting for one to leap out of a tree.

Slow-moving risks that build over a long period of time seem like they should be easy enough for us to step out of the way, but unlike fast-moving risks, the slow ones do a surprisingly good job at knocking us off our feet. A slow change in the price of gas over several months may stir little response from us, but a sudden jump to four or five dollars per gallon knocks our socks off. If a skunk is a hundred feet away and creeps toward you at four inches per hour, you probably won't panic . . . until perhaps it's too late.

One problem we have with slow-moving risks is that our eyes and minds respond strongly to quickly moving stimuli, whereas slow movers often don't stir much of a response. It's easy to see the second hand move on a clock, but it's almost impossible to see the hour hand move. You know it's moving, and if you look away and then back again fifteen to thirty minutes later, you will see the difference; but if you stare at it continuously, it's hard to say at any given moment that you see it moving. You might have tried this experiment in high school, watching the clock on the wall, waiting for the hour hand to move, convinced it was broken.

There's an old story about a frog being placed in a pot of water while the water is slowly heated. If the water is hot when the frog goes in, the frog will jump right out, the story goes. However, if the water is heated slowly enough, the frog never senses the risk and stays in the water until it's too late and the frog gets slowly cooked. I've also heard that the story isn't true, for frogs at least; frogs are smart enough to get out of the water when it gets too hot. But the story does seem to describe human behavior pretty well. Despite our myriad amazing abilities, we often find ourselves caught off guard by slow-moving risks.

Evolution does a good job at giving us good responses to immediate threats, but it does a much poorer job at providing effective responses to risks that develop slowly, are far away, or are spread out over a large area. The evolutionary impact of immediate threats is greatest because of the direct link between responding and surviving. To survive a leaping cougar, you must act immediately—or else. There's not a lot of time for thinking it through. Evolution is a harsh teacher, priming us to act quickly to survive threats like this, but we also have a blind spot for risks that fail to hit these hot buttons. A slow risk that develops over years rather than minutes does not trigger the same responses. For instance, if I pick up the phone and a greasy, tasty pizza shows up right away, and there's no risk apparent, I'll quickly learn to call for pizza whenever I'm hungry. That's how it seems to go for many college students, particularly when using their parents' credit cards. If you pick up the phone and get an electric shock every time you try dialing for a pizza, you would stop picking up the phone. If the pizza, the electric shock, or the heart attack accompanied with poor diet show up twenty years later, you won't learn much—the time between the action and the feedback is so great that the connection between cause and effect is not made in our mind.

We still encounter immediate threats in the modern world, like when we have to leap aside to avoid speeding cars instead of charging buffalo. But the modern world is also full of slow-motion hazards that take decades to overtake us, risks that we are poorly suited to deal with. We fail to save for retirement, ignore the threat of climate change, miss a looming financial crisis, and deny the risk of heart disease because these risks creep up on us slowly over time. Our inability to see these risks coming makes them all the more dangerous.

In movies, environmental hazards come in the form of dramatic, momentous, and immediate cataclysms. The 2004 climate change movie *The Day after Tomorrow* did not feature melting icecaps, receding glaciers, or slowly rising sea levels, but instead depicted tornadoes ripping apart Los Angeles and the overnight freezing of one-half of the Northern Hemisphere while a climatologist played by Dennis Quaid tried to save the world and his son. He successfully saves his son but not the world, unfortunately. The long-term consequences of climate change may be dramatic but they are not likely to happen overnight, like in the film. Even if the long-term impact of slowly rising sea levels would be dramatic, submerging cities and changing life as we know it, telling the story in terms of a slow change over many years would not sell as many movie tickets and often fails to hold our attention outside theaters, as well. A cataclysm, on the other hand, really grabs our attention and grabs it hard.

Environmental risks often develop slowly, taking years and

decades to build up. Pollution, resource depletion, overfishing, habitat loss, and extinction do not happen overnight. It has taken decades or longer for these environmental challenges to reach their current states, and it will take a long-term perspective to reverse them, but it is a long-term perspective that does not come easily to us. If water pollution accumulates slowly over a period of many years, gradually eroding the surrounding ecosystems with heavy metals, solvents, or oil, people might not pay all that much attention. It has happened many times. A cataclysm like a flaming river is another story, though.

When the Cuyahoga River in Ohio caught fire in 1969, it wasn't the first time.[2] The river had countless tons of raw industrial waste dumped into it for decades and had accumulated a thick, oily layer on top with practically no life found in its waters. The slow deterioration of the river had stirred little response over the years until the river caught fire in 1969 and the shocking image of a flaming river jolted the country into action, helping along the birth of the modern environmental movement and the passage of the Clean Water Act. Polluted waterways were a dime a dozen at that time, but you don't see a flaming river every day. If not for the shocking image of the fire and the national attention it gained, the urgency to take action may have taken much longer to arrive, if it did at all. We, as humans, often delay taking action on a problem like this, putting it off unless a crisis forces our hand.

A FAILURE TO COMMUNICATE?

We respond well to risks that are here and now, easy to see, and easy for us to fix or avoid. However, in addition to happening slowly, environmental problems also develop over large areas and with diffuse responsibility, further buffering the sense of accountability and risk. It's not that one person is responsible for air pollution or water pollution. These problems are regional, though, and can often be seen, felt, and addressed by people and governments in that region. But climate change is on a whole different level.

Climate change does not have one person who is responsible, or even just one country. It is a global problem, compounding the difficulty of taking action. Even if people in one city or one country take on the problem, they can't solve it on their own. And the long periods of time involved in climate change make it hard for us to see the connections between environmental reactions and our actions, or to see the necessity for immediate action to deal with any risks.

While weather may change overnight and it is clear to all of us when we look out the window, the climate takes years and decades to change, or even longer. A change in the climate can even take place over centuries and still be rapid in relation to earth's history of billions of years. But in human terms, a change like this, occurring over the course of multiple generations, might be practically imperceptible.

According to data from NASA, the global average temperature is rising about 0.29 degrees Fahrenheit per decade; since the 1970s, the earth's average surface temperature has risen by about one degree Fahrenheit.[3] For comparison, consider that your average location in a temperate climate will swing by as much as twenty or thirty degrees from day to night. The data about climate change does not depend on one group's approach or data—studies looking at other data from other sources reach the same conclusions. For those who question whether climate change is real, I can't review all the scientific data here because that would be a different book. And even if I did review the data, you would not be swayed, as we shall see.

A temperature change of one degree Fahrenheit might not sound impressive, but the change is not evenly distributed, and the global impact of climate change may accelerate, as many scientists think is likely, leading to more extreme weather. In addition, the impact of climate change is not just about temperature but also about things like rising sea levels.

The sea level is already rising about three millimeters a year, and is likely to rise even more as the melting in Greenland and west Antarctica accelerates.[4] Once again, the pace of change isn't enough at any given moment to get the majority of the world's people up in arms agitating for the reduction in greenhouse-gas emissions, but over a period of decades, a few millimeters a year adds up to a meter or more of sea-level rise, enough to submerge a good-sized chunk of Florida, Manhattan, Shanghai, San Francisco Bay, Venice (both in Italy and the beach in Los Angeles), half of Bangladesh, and plenty of other coastal regions around the world where hundreds of millions of people live.[5] The long-term impact of such a change could hardly be greater; but, as of today, these threats don't have many people marching in the streets, at least not where I live. When people go to bed at night and get out of bed in the morning, they might be worried about their kids, their health, their jobs, their bills, and their homes, but few probably wake up with climate change as their most urgent concern for the day.

Global talks about climate change have been going on for many years, but so far there's no significant, great global effort underway to divert us from a rapid slide downward. The rate of greenhouse-gas emissions rose faster than ever in 2010, after briefly dipping earlier in what many call the "Great Recession" of 2008.[6] And international negotiations on a treaty after the Kyoto Protocol of 1997 are making little progress. So far we're doing little or nothing to stop climate change in a concerted global way. There are exceptions to this, with many individuals, cities, states, and even some nations mobilizing to take on this challenge, but the response is nowhere near what would be needed to actually make a dent in this global problem. Not even close.

There are many reasons why we don't do more to respond to this problem. You might be thinking that the problem isn't real, but if so, you're in the minority. Another explanation is that we don't know about climate change or we don't understand it. Or maybe we know but we don't really care. None of these "reasons" seem to be the problem though.

Depending on the surveys you look at, most people know about climate change and even agree with the statement that humans are causing it. Anthony Leiserowitz, director of the Yale Project on Climate Change Communication, writes, "Since the year 2000, numerous public opinion polls demonstrate that large majorities of Americans are aware of global warming (92%), believe that global warming is real and already underway (74%), believe that there is a scientific consensus on the reality of climate change (61%), and already view climate change as a somewhat to very serious problem (76%)."[7] The numbers vary from survey to survey, but the conclusions are always the same. "At the same time, however, Americans continue to regard both the environment and climate change as relatively low national priorities." We know there's a problem. We're just not excited about doing something to fix it.

LIVING IN A BUBBLE

The amount of data being generated about climate change and the amount it is discussed in the media have gone up even as our global response has stayed flat. In fact, while information about climate change has increased and scientific data continues to pile up, the perceived risk of climate change has decreased, according to surveys. In Norway, the percentage of people who were very worried about climate change dropped from 40 percent in 1989 to 10 percent in 2001.[8] You would think that providing more information about this massive global risk would change minds and mobilize more of us to tackle the truly enormous task of fixing it. Scaring people with words seems to have exactly the opposite effect and instead is driving people further away from change. The more data that is heaped onto the public, the further we seem to drift from action, even steeling ourselves against it.

If information is not the problem, then perhaps the answer is that we need to experience climate change ourselves in order to really react. Experience is a powerful teacher, and very often we can't really believe something just by reading about it. With harsh winters in the United States in 2008 and 2009, many people concluded that climate change was a hoax; the snow and ice they saw with their own eyes carried more weight than scientific graphs and models, leading people to say often, "What global warming? Look how cold it is outside!" But even when we do see the evidence for climate change with our own eyes, we have not leapt into action. In Norway, winters have become conspicuously milder, with less snow over the last several years. The change is obvious to inhabitants there, as described by sociologist Dr. Kari Norgaard of Whitman College in her book, *Living in Denial*, but the response to the change has been less than obvious.

Norgaard spent almost a full year living in a small town in Norway, which she calls "Bygdaby" in the book. The people in Bygdaby are well educated, politically active, and care a great deal about others around them in their tight-knit local community, in their reserved

Norwegian way. Norway is a wealthy country and one with a very egalitarian social system. "Bygdaby" has about ten thousand inhabitants, many of them farmers and many of them making at least some of their money from winter activities such as skiing and ice fishing. The average temperature in the winter there has become obviously warmer over the recent years, delaying the first snows of winter from November to January. In fact, in the winter of 2000–2001, the late snow and rain caused flooding. The change in weather was linked, both in the media and in the discussion of people living there, to climate change.

And yet despite all of this, when Norgaard talked with them in a variety of settings, the people consistently avoided the topic of climate change. The more Norgaard got to know Norway and its people, the more she came to see that a confluence of influences was shaping their feelings about climate change. They know about climate change, and they are disturbed by it. To some, Norway is even responsible, as much of its wealth in recent years has stemmed from the sale of its rich oil supply that makes Norway one of the world's largest oil exporters. But talk about climate change and how to fight it are almost nonexistent there. At first glance, it seems like something doesn't fit.

Norway and Bygdaby are safe and prosperous places. People work together and talk about local issues. They know that climate change is already having an impact and that the impact will grow, but they avoid talking about it directly. Bringing it up leads to awkward silences. This is because the citizens of Bygdaby want to maintain in their mind the safe world they live in (or at least their vision of a safe place), and the way to do this is through denial. Trying to talk about climate change feels like a threat to them, more of a threat than climate change itself.

In the back of the Norwegians' minds, submerged, is a pervasive fear of the threat of climate change. Climate change is reshaping our planet all around us, and not for the better. It throws a pall of uncertainty over many things we take for granted today, including the type of world our kids will live in. A problem like this can make you or anyone else nervous. When you're talking about cities underwater, more and more extreme storms, a lack of drinking water, agricultural stresses, disease, war, and millions of refugees—this is very bad news indeed, and not just "uh-oh" nervous, but waking-up-in-a-cold-sweat kind of nervous.

When dealing with anxiety on this scale, people can do one of three things. They can (1) take action to fix the problem, (2) stay anxious and sweat it out for a long, long time, or (3) do a quick mental sidestep to turn off the anxiety and make themselves feel better. The simplest, fastest, and easiest of these options is usually the third. For the majority of people who feel that climate change is a risk, they also believe that there's not much they can do about it, so they push it from their mind. As of right now, the cities are not yet submerged and probably will still be above water next week; add to that the fact that we've got plenty of other stuff to worry about for now, so what are we going to do?

"Climate change is disturbing," says Norgaard. "It's something we don't want to think about. So what we do in our everyday lives is create a world where it's not there, and keep it distant."[9]

The inhabitants of Bygdaby, and all of Norway, for that matter, are not the only ones keeping climate change as far out of their mind as possible. People in the United States, for example, have many of the same perspectives. The view of many is not that climate change is the threat but that the talk of climate change is the threat, and denial feels like a quick solution. The reason providing more information has a perverse impact on taking action is that the more people learn, the worse climate change sounds, and the greater the potential is for anxiety. The greater the anxiety, the more attractive that denial becomes. Rather than scaring people into action, heaping on more information only leads to more strenuous denial.

"It is this quality of angst, a condition that social psychologists tell us we are profoundly motivated to avoid, that essentially makes climate change 'unthinkable,'" says Norgaard.[10]

EXPERIENCE IS THE BEST TEACHER

One way this might change is through the power of personal experience of climate change, feeling real pain. "In most [W]estern countries, people lack personal experience of climate change, which is considered to have direct impacts on people's lives only in far-away places or the distant future," said Elke Weber.[11] As they say, experience is the best teacher, and sometimes the only teacher, with floodwater rising around your feet having much more of an impact than any pile of scientific data, no matter how high the pile is.

Severe flooding has struck the United Kingdom on several occasions over the last decade, in 1998, 2000, 2004, 2007, and again in 2009. Hundreds of thousands of people were directly affected by these floods, and the nation as a whole was seriously impacted by a disruption of daily life, a story that was carried extensively on the nightly news throughout the affected regions. When the floods occurred, the media clearly linked the unusual and repeated flooding to climate change, saying that increased and extreme rainfall were predicted by climate change models to occur in the United Kingdom exactly as was being seen.

While at the University of Nottingham in 2010, Alexa Spence

and her colleagues looked at how experiences with flooding have changed perceptions of climate change in the United Kingdom and the willingness to do something about it.[12] Whereas even those who profess to worry about climate change often fail to act, believing it won't make a difference, those in the United Kingdom who had lived through recent floods believed that their actions would make a difference. Personal experience of something big like flooding has the power to make climate change undeniable and jolt us into action. Spence and her colleagues concluded, "We show that those who report experience of flooding express more concern over climate change, see it as less uncertain and feel more confident that their actions will have an effect on climate change."[13] Seeing climate change in action in your own life and feeling its affects directly might be what it takes to get things going in your community.

A 2010 US Forest Service report examined climate change in Alaska, which has already been impacted by climate change and will see an even greater impact in the future. "Over the past 50 years, Alaska has warmed at more than twice the rate of the rest of the United States."[14] Alaska has seen winter temperatures go up by more than six degrees Fahrenheit and now has ten fewer days of snow than it once did. All around the Arctic, the permafrost (land once permanently frozen in areas like Alaska, Canada, Scandinavia, and Russia) is melting, turning the ground from frozen solid to mud. The melting permafrost is destabilizing buildings and roads and may threaten the Alaska Pipeline.

These changes are obvious enough that ordinary people are taking notice. In Alaska, over 43 percent of those surveyed are completely convinced that global warming is happening, which is significantly more than people in the rest of the United States, where only 23 percent responded with the same level of conviction.[15] Florida has experienced changing weather and will be heavily impacted by rising sea levels. This is reflected in a trend in surveys, where more people in Florida than in the rest of the United Stated say they want to take action.[16]

Water rising around your home and around your feet makes climate change much harder to deny. Even now the signs of climate change continue to emerge all around us. In February and March of 2012, weather across the United States was unusually warm, thirty to forty degrees higher than normal in some areas. There were tornadoes in Michigan in what would normally still be winter. Mosquito populations are up, and maple-syrup production is down. For many, it's nice to have a warm spring, strolling in New York City in shorts in February, but these changes come with a price . . .

And things are just getting started. Back in Alaska and all around the Arctic, the melting permafrost holds an even greater threat, one that is starting to bubble up as the permafrost thaws in the form of methane. Methane is a potent greenhouse gas that comes from rotting plants, like those that have been frozen in the permafrost for tens of thousands of years. As the permafrost thaws, the plants rot, and methane starts to bubble up. Even now, lakes around the Arctic are starting to bubble with methane, with far more waiting to rise to the surface. With methane twenty times more potent at trapping heat than carbon dioxide, this methane could make any efforts to block climate change much, much harder. At some point, we might find ourselves on a steep downhill slope, slipping toward the edge with no power to change what has already begun. It's hard to say where that edge is exactly, but we'll only know for sure long after we've crossed it, which would be a very bad thing.[17]

PAYING TOMORROW

One factor that keeps us from fighting climate change is that the solutions often require us to start paying today for rewards that mostly arrive decades from now, a combination we are inherently bad at dealing with. We like to buy now and pay later, not the opposite.

We might raise the tax on gasoline, or make coal-fired power plants pay to capture their carbon dioxide and put it somewhere other than in the air, for example. Either of these would provide an economic incentive to reduce greenhouse emissions, the kind of incentive that would drive us toward greener and cleaner long-term solutions. This is good not just for polar bears but also for people; in the long run, there would be an enormous economic benefit to fighting climate change. Failure to act will eventually cost us trillions of dollars, with as much as 20 percent of global gross domestic product (GDP) potentially lost to storms, disease, famine, pollution, and flooding.[18] (To put that into perspective, in the global financial recession of 2008–2010, the United States lost about 5.1 percent of GDP, at its worst.[19]) And 20 percent loss of GDP due to climate change is a conservative estimate, assuming our delicately intertwined global economy doesn't come completely unglued under the stress of rising oceans, rising populations, and declining food and water. But still, even if we know all of this, most of us today don't sense imminent peril breathing down our necks or rising around our feet, so we don't want to pay to fix it today either.

A bird in the hand is worth two in the bush, as the saying goes, capturing perfectly how we perceive the cost and benefit of short-term versus long-term risks. We'd much rather eat our bird now than have a whole flock of birds a few decades from now. In the case of climate change, we're asked to spend slightly more on energy today to protect future generations against much greater costs from extreme weather, rising sea levels, and increased temperatures. What many of us think about that is, "You want my money for what? I don't see that, but I do see a pile of bills I need to pay." Spending money to fight climate change might make sense from an analytical perspective, which is why major insurers recognize the risk and the importance of fighting it. On an emotional level—the place where we spend most of our time—we would rather spend our money on stuff we need or want to buy today.

While the future is far off and uncertain, it's easy for the more concrete present to dominate our thinking and to be valued far more highly. "People often apply sharp discounts to costs or benefits that will occur in the future (e.g., a year from now) relative to experiencing them immediately," the American Psychological Association (APA) wrote in the report *Psychology and Global Climate Change*, addressing why we do so little to address climate change, with future benefits spread across decades, despite the enormity of the problem.[20]

Elke Weber, director of the Center for Research on Environmental Decisions at Columbia University looked at humans' tendency to discount future benefits by telling volunteers in a study that they could either win a fifty-dollar Amazon.com gift certificate that day or potentially receive a larger gift certificate later.[21] Another group was told that they could win a gift certificate worth seventy-five dollars in three months or get a smaller gift certificate that day. Either way the question was asked, the result was the same: we want it now. The subjects valued immediate rewards much more than later rewards, far more than any financial analysis could justify. We view these two scenarios differently because in one case we are asked to delay consumption, while in the other we are offered the opportunity to accelerate consumption. We like accelerating consumption.

We don't like to wait, and we like it even less when asked to wait longer because it feels like something is being taken away from us. It's all a matter of how the question is presented and how volunteers work through the problem in their minds, breaking it down based on impatient thoughts (e.g., "I want it now") and patient thoughts, delaying gratification.

It is reminiscent of a story about a farmer with a hole in the roof of his barn.

> On a sunny day, the farmer's friend, who lived on a neighboring farm, came for a visit and they took a look at the barn.
>
> "You really should fix that," his neighbor told him, pointing up at the hole.
>
> "I don't have to fix it now—look, it's sunny out," the farmer said. Later, on a day when it was raining hard, his neighbor came by again in his raincoat and asked once more about the roof, as water poured through the hole and onto the hay and animals.
>
> "Well, I can't fix it now—it's raining," the farmer said.

Somehow there's never a right time for fixing the barn—or for fixing climate change.

There was probably a time in human prehistory when taking a short-term perspective on life made sense. Life was short, and the threats against it were constant and immediate. Avoiding predators and ensuring food were immediate concerns; you would be lucky to survive long enough to worry anything beyond next week. "A focus on the here and now probably may well have short-term survival and some evolutionary advantages, especially in the simpler environments we lived in when our current processing patterns got shaped," said Weber.[22] The times have changed though, while we haven't. We're still on the lookout for immediate risks acting directly on us, and we're blind to long-term risks, like climate change, with far greater long-term costs.

THE WHITE-MALE EFFECT

It's no secret that not everybody feels that climate change is real. I've heard the word *hoax* tossed around, for example, with some arguing that the science or the facts are uncertain, that "Climategate" casts a pall over the whole thing. Rather than being based in the science or the facts, the differing perceptions of climate change may really stem from different worldviews. Not too surprisingly, we don't all see the world the same way, including its many risks.

As social creatures, we tend to share our view of the world with other likeminded people, becoming even more likeminded as we gather in groups. The more we share of our worldview, the more this influences our view of the world's risks as well. Dan Kahan at Yale University, as well as others, calls this impact of groups on how we see the world and its risks *cultural cognition*. "People endorse whichever position reinforces their connection to others with whom they share important commitments," writes Kahan. "As a result, public debate about science is strikingly polarized. The same groups who disagree on 'cultural issues'—abortion, same-sex marriage and school prayer—also disagree on whether climate change is real and on whether underground disposal of nuclear waste is safe."[23] There doesn't seem to be a logical connection tying together such varied topics, but there doesn't have to be if they're based on culture.

According to this idea, there are two main cultural groups when it comes to questions of climate change. One group includes people sometimes called "hierarchical individualists," who value structure, individualism, and business opportunity, and are skeptical of health risks, technology risks, and environmental risks. Oftentimes, the discussion about climate change is structured in a way that this group finds threatening to their culture, such as the idea of restricting our choice of cars in order to improve fuel efficiency or the idea of instituting carbon taxes to reduce greenhouse-gas emissions. And when we identify with a set of beliefs, with a culture, then a threat to this culture feels like a threat to us. The real issue from this perspective isn't so much about the facts regarding climate change but about the shared culture those in the group are defending. "Citizens who hold hierarchical and individualistic values discount scientific information about climate change in part because they associate the issue with antagonism to commerce and industry," wrote Kahan and his coauthors. Many of those in the hierarchical-individualist group are white males, including some of the leading climate change deniers, leading to this phenomena sometimes being called the "white-male effect." There is no doubt that this is a simplification and the group includes many who are not white males but who still maintain this worldview. A more neutral label like *hierarchical individualist* might be more accurate and less likely to provoke a fight, but it doesn't roll off the tongue quite as easily. [24]

The other cultural group can be called "egalitarian communitarian," a group that feels environmental risks, as well as health and technology risks, are great. This group emphasizes the idea that "we're all in this together," so to say, so it tends to be more accepting of laws and regulations that emphasize these values. I've got a feeling that this group also has its own set of cultural beliefs that egalitarian communitarians cling to tenaciously and act to defend, which flavors how they see risks. The reason the discussions between these groups seem so unproductive is because they aren't speaking the same language. Arguing facts against culture is like responding to a math test with a painting.

Our need to belong (see ch. 8) is behind this failure, driving us to think, talk, and act together, engaging in what some call "groupthink." People are reluctant to step out of line if they feel this will get them booted from the group, so sometimes our thoughts are not our own. Even our much-vaunted intellectual capabilities packed in our big brain are not immune to groupthink. While we often view our feelings and our intellect as two separate things (as I describe them in this book on occasion), they are actually intertwined. Our intellect does not function independently of emotion; we are not Spock or a computer. Instead, since we're human, we often employ our intellect in what is sometimes called "motivated reasoning" in the service of our feelings about things. We reach conclusions first, based on what the group and our emotions decide, and then do some backfilling with intellect to provide support for the decision.

Viewed this way, intellect is not a dispassionate observer but the henchmen for our feelings. Our intellect builds a case, helping us to believe what we already believe, like the latest word from our favorite climate change deniers, and goes to work chopping up those with whom we disagree, cutting them down to size.

A LOSING ARGUMENT

In addition to the previously cited reasons why we don't do much about climate change, another is that the data from scientists is, well, scientific data, and this isn't something most of us can easily relate to. To many people, scientific arguments sound like Charlie Brown's teacher—"Wa wa, wa wawawaa..."—gibberish in a complex, irrelevant language that means nothing to us. The APA found that while those saying we need to fight climate change rely on scientific data, fear and anger are used to argue that climate change is a sham being used to take our money and jobs. When the discussion pits values against data, data will lose if the values resonate with us.

Eric Corey Freed is a green architect and an author living in the San Francisco Bay area and lecturing all over the country about climate change. He talks to an estimated ten to fifteen thousand people a year, including everyone from hardcore greenies to vehement climate change deniers, and he has tried a variety of approaches to motivate his audiences toward action. At one time, he would try arguing his case with facts, but he found this was for the most part unsuccessful. Those in the audience who already agreed would nod their heads knowingly in approval, but those who did not agree were seldom swayed by a barrage of facts and figures.

"We don't really want to change our viewpoints, and most people can't change your viewpoints for you," said Freed in an interview.[25] "We solicit an emotional reaction from people, then we fight back with comments going back and forth."

While change is possible, whether it's changing our mind about climate change or about adopting a healthier diet, it's rare for people to be *persuaded* into changing their mind because no amount of arguing, facts, or scare tactics will change us. "All change has to come from you. Even the people closest to you, even your spouse, won't change you, though they'll probably try. Real change comes from you, saying 'You know what, I'm sick of this.' It comes after years of introspection until [you] get [to] a point where you say 'I'm going to change,' whether it's losing weight or fighting climate change."[26] But usually we don't do this. Maybe this relates to our cultural cognition at work as well.

"We like greasy food, watching TV, and being a couch potato," said Freed. "I like it too. If I tell someone 'You're going to get off your couch, live in a community that's walkable, and bike to work,' it sounds awful." We seldom like it when people tell us what to do, particularly when it sounds difficult and time consuming, and especially when we know they are correct.

So with all this in mind, how will we finally address the risk of climate change? "My viewpoint has changed from 'We're going to educate everyone' to now thinking, essentially, that things will get worse and worse until everybody is forced to do something. I think that we are not going to stop it. It's too late, so let it happen, and then we'll have a little 'I told you so' moment as people scramble to survive."[27]

At this point while talking to Eric, I got a little depressed, in part because I've been thinking the same thing myself. It feels like there are seven billion of us all on this runaway train headed for the cliff with the tracks blown out, but unlike in the movies, I'm afraid there's nobody who's going to jump from a horse to the train and pull the brakes. There are brave souls doing their best, but so far nobody's really figured it out. And meanwhile, most of us are doing our best to avoid seeing the cliff. I thanked Eric, and tried to sit back and enjoy the train ride.

But sometimes a crisis is the only thing that gets us to stand up and take urgent action. Necessity is truly the mother of invention, and just about everything else. This isn't true just for climate change but also for many things. When are people most likely to improve their habits to avoid heart attacks? After they've already had one and they feel like they have no other choice. Even then it can be hard getting someone to sit up and really see the dangers he or she faces and to actually convince him or her to change lifestyles.

Our failure to fight climate change is often blamed on climate change deniers, who question whether climate change is a real problem. Exxon Mobil, the Koch Brothers, and others have been well documented to have carried out a campaign to create doubt about climate change in order to benefit those with financial interests in coal and oil. Really it's not this simple, though. While there has been a highly effective campaign to sow doubt in the American public about climate change, those leading the campaign of denial are not the only ones involved. The climate change deniers are not alone; it takes two to tango, or maybe a few billion of us. The success of the climate deniers stems from our own innate programming that makes us receptive to their message. If we don't see an immediate threat and can't easily fix it right here and right now, it's in our nature to push the threat out our mind, to deny that it's even a problem at all. Even those who believe in climate change and say it's caused by humans are not all clamoring for more effective global action. Climate deniers don't invent how we think; they merely exploit our own desire to create a safe bubble of denial, to build a levee holding back the rising tide of uncertainty. It seems that there is a little denier in all of us, and our inner denier comes out not just to deny climate change but also to deny many other risks as well.

A persistent, common, and fundamental error we make is that we tend to be overly optimistic about our own personal fate, even when we are pessimistic about the fates of others. We also do our best to create an optimistic view of our personal prospects in the world, viewing our world as safe and friendly. Most of us spend copious amounts of our time thinking that bad things will never happen to us. Maybe this belief is essential to lead a normal life. The idea of climate change runs counter to this rosy worldview, but individuals do not feel a ready way to reverse climate change, one in which an individual's efforts quickly lead to visible results, so it is easier to mentally avoid it.

THE BRIGHTER SIDE OF BAD FOOD

Climate change is not the only risk we deny. We deny long-term health risks like heart disease and diabetes for similar reasons. Despite a broad range of medical advances, heart disease kills one in five people in the United States, making it one of the biggest risks we each face.[28] The deterioration of our arteries is so slow, advancing gradually over decades deep in our bodies until the artery feeding the heart is blocked, that we sense little hazard in our daily lives. Even if we know intellectually that heart disease is a significant long-term threat because we've seen the numbers, even if our doctors tell us that we really need to change our ways or else, heart disease seldom feels like an immediate threat that we must act on today. We're not likely to have a heart attack today, and we know it.

Why don't we take action? The solution is pretty simple, really. Eat less bad food, eat more good food, and exercise more. If it's so simple, why do we find it so hard?

A big part of the problem is that bad food can taste pretty good and exercise can be pretty hard. Fast-food restaurants, for example, know how to tap into this, hooking us with the taste of pure sugar and fat that our ancient genes crave. A seventeen-year-old in the United Kingdom recently checked into the hospital after a lifetime of eating almost nothing but Chicken McNuggets.[29] It was all that she liked to eat, she claimed. She was not a healthy person.

To be honest, I like Chicken McNuggets too, once in a while, particularly with the sweet-and-sour sauce. The UK girl and I are not the only ones. Having a taste for bacon, fat, and sugar is not a new thing. Fat and sugar are loaded with the kind of highly concentrated calories that could save your life if you were to live in a world with limited food, the kind of world our ancestors lived in. There were probably feast times when fruit was in season or a hunt was successful and there was more food than they could eat. The feast times would then be followed by plenty of lean times or even famine. When this happened, it made sense to bank as many calories as you could, laying down reserves since your next meal was uncertain. And it makes sense that evolution would prime us with tastes and behaviors to lean in that direction, to eat, drink, and be merry, avoiding starvation. We are wired to be eating machines because of the world we evolved in.

The world has changed, though. While there are people in the world who go to bed hungry, a growing number of us live in an environment with immense quantities of food available for just pennies. A can of soda contains thirty grams of sugar, or 140 calories. A Double Gulp, an enormous bucket of soda big enough to fit on your head (perhaps I exaggerate here), contains sixty-four ounces of sugar, about a cup's worth. It's hard to imagine that we would willingly sit down and eat a cup of sugar, but for a growing number of people, sodas are the biggest source of calories in their diets. We might not eat that much sugar with a spoon, but many of us will drink it from a giant cup with no problem.

Faced with food everywhere, our inner feeding machine is running amok, and the off switch is hard to find. Our ancient hormones and neuronal circuits that kept humans from starving still create such a powerful drive to eat that we're eating ourselves to death in a global epidemic of obesity that comes with a side of heart disease, diabetes, and high blood pressure. While we have the ability to choose to eat less, and while there are many inspiring examples of individuals who win the battle with their hormones to improve their health and their lives, the statistics say that far more people are losing the battle. It sometimes seems like we are slaves to our ancient programming.

Even when our doctors tell us we have a serious risk of dying of heart disease, the responses are similar to those we have when facing climate change: do nothing and worry a lot, deny the problem, look for quick and easy answers, or look for long-term solutions that might not be easy (like diet and exercise). Denial is a popular option for heart disease, just like climate change. Even if their doctors point out the risks, many people will choose not to listen. Knowing about the long-term threat, we become anxious about it; if the anxiety is maintained for a long time, day after day, it starts to feel like more of a threat than the heart disease. So we have a choice to make. One option is to change our lives, improving our diets and exercising regularly. But the lure of cheesecake today (the second option) is often greater than the distant threat of a heart attack years from now. Changing our diets, changing our lifestyles, and exercising routinely take years of effort, while denial relieves concerns right away with virtually no effort at all.

For a study in the United Kingdom, 381 patients judged by their doctors to be at high risk of heart attack were surveyed. In this population, three-quarters were obese, over one-half were diabetic, and 57 percent had high blood pressure even with common usage of drugs to treat the condition.[30] All of these conditions greatly raise the risk of a heart attack, but we do our best to not think about it. Even after strong warnings from their doctors, two-thirds of the UK patients in the high-risk heart-attack group convinced themselves that they didn't have a problem, that they had no more risk than anyone else their age. And then having convinced themselves of this and seeing no problem, they failed to improve their diets or exercise.

Studies show that many people don't call the doctor even when they suffer chest pains because they refuse to believe that they could have a heart attack. They attribute their chest pains to indigestion, telling themselves that it will go away. The pain must truly be severe before people will admit to themselves or anyone else that they're having a heart attack, doing their best to deny the risk even as it happens to them. When this happens, denial can be lethal; at a time when getting medical attention sooner rather than later can save your life, delaying treatment can mean death.

Smoking is a different kind of inconvenient truth. Despite extensive public information campaigns in the United States, 20 percent of Americans persist in smoking, and the number may be increasing again. There is more to this persistence than an addiction to nicotine. People also keep smoking because the threat of cancer doesn't feel real and immediate to them. When we persist in risky behavior, like smoking, we often deny the risk, using self-exempting beliefs where we acknowledge the risk to everyone else but employ a mental defense that works just for us, believing that we don't smoke enough to be at risk, or that the way we smoke reduces the risk, or any number of bogus reasonings we can come up with in order to exempt ourselves.[31]

Why do we deny risks like climate change and heart disease? Because denial often works, in its own way. It may not make the problem go away in the long run, but it helps us to feel better, and this alone can have surprising benefits. A fifteen-year study of self-rated cardiovascular risk by Dr. Robert Gramling and his colleagues looked to see how the viewpoint people had about their own risks affected their actual risks. Over 2,800 people in New England took a survey in the early 1990s and then followed up fifteen years later to see how they fared over time. The survey asked men and women to judge their own risk of cardiovascular disease as high, medium, or low. Men who ranked themselves as low risk fared three times better than those who judged themselves as high risk, even when the data was corrected for actual risk according to various physical measures. Feeling that they were low risk, they became low risk; their denial of risk seemed to pay off by actually reducing their risk. (Curiously, the effect seemed large for men but could not be detected for women.) It's possible that optimistic people are simply healthier, doing healthier things like exercising. But even when the usual cardiovascular risk factors in these patients, like LDL-cholesterol or the inflammatory marker CRP (c-reactive protein), are corrected for, optimism still shields these patients against the risk of heart attack.[32]

Denial can also help people recover from a heart attack after they've already had one. Researcher Jacob Levine interviewed men who had suffered a heart attack and were hospitalized, either recovering or awaiting bypass surgery. You might think that heart disease would be impossible to deny at this point, but you might be wrong. It's hard to overestimate the power of denial, at least in the short-term. A large number of the patients had little interest in the nature of their disease, actively avoiding the question and avoiding follow-up treatments to improve their condition. After the interview, the patients were followed for a year to see how they fared. While in the hospital, those who were the most in denial actually fared better at first, spending fewer days in the hospital and being discharged sooner. Optimistic people might be in denial about their risks, but sometimes their optimism pays off.[33]

There is, however, still a down side to denial. Levine also found that high deniers ended up spending more time in the hospital in the long run. Their denial helped them to bounce back faster initially, probably by avoiding further stress at a time when stress on their hearts is very bad, but denial also caused them to avoid changing behaviors like diet and exercise that make a big difference in one's cardiovascular health over the long run. Positive thinking truly is powerful and often beneficial regarding our health, certainly better than excessive worrying and stress. But this optimism can also lead to the denial that a problem really exists, and it can make things worse in the long run, as we saw with climate change, as well.

DENYING THE OBVIOUS

Perhaps denial is an adaptation we have developed in order to protect ourselves from stress and the health risks that go along with it. If denial lowers stress and the chronic health problems it creates, it has clear benefits. When risks are present over a long period of time, being in denial is like taking a vacation, just without all the trouble and expense.

Meanwhile, the impacts of climate change are not all in the distant future. Increasingly severe weather caused by climate change is already happening, as seen in the floods in the United Kingdom. Animal and plant species around the globe are already stressed by changing seasonal patterns. Growing seasons are disrupted, floods have increased, droughts have increased, and whole forests are getting wiped out by drought and insects. These events are not restricted to the distant future, like the year 2050 or the year 2100. They're already happening today, and not just for polar bears.

In the tiny Cartaret Islands located far out in the Pacific, the islands' 1,700 inhabitants are leaving, relocating to another island about eighty kilometers away. The islands have been inhabited for a thousand years, with people living off of fish and the simple crops they could grow there. Today, life on the islands has become impossible. The seven island atolls are losing the fight with the sea, with drinking water becoming salty and killing the ability of islanders to grow anything but coconuts. Now even the coconut trees are dying. Storms have been chipping away at the islands, and they're expected to be underwater by 2015. These island inhabitants may be the world's first climate change refugees, but they won't be the last. Other islands in the Pacific are expected to be next. "It is not just us, though," said Ursula Rakova, a former Carteret islander. "It is going to happen to others right through the Pacific and in other parts of the world. The pollution has to stop." As the problem continues to grow, it affects everybody on the planet, including you and me.[34]

One result might be that increasingly obvious and urgent consequences of climate change may force us into action. In the 1960s the ignition of the Cuyahoga River stirred national outrage and led to the Clean Water Act. Perhaps it will take another Cuyahoga River–style crisis to shock us sufficiently to mobilize a global response to climate change.

Experience is a great teacher and motivator, but sometimes the lesson comes too late. If you need to have a heart attack before you start exercising, this is a very hard lesson to learn, and for some people there won't be a second chance. If you have to wait for climate change to become undeniable through ever more urgent crises hitting one after another, then the lesson is coming quite late in the game. If preventing climate change is difficult, then stopping the climate from changing after it has already changed may prove impossible, even if we're motivated. It's a lot harder to fix the leak in your roof when it's already raining. Or when your roof is underwater.

There are some ways out of this before the seawater is lapping at our feet. The power of anxiety to drive us into the arms of denial means that ratcheting up the rhetoric with ever scarier scenarios will only drive most of us further in the wrong direction. Maybe we can get around the culture wars by finding issues and values we can agree upon rather than focusing on the ones we don't. And the way we can do this is by talking about values rather than facts. Another clue might come from those who study disaster preparation, suggesting that we should make climate change action social or take the Coke approach to marketing it (see ch. 4). Telling the story of specific people rather than talking about the future and global statistics helps us to stop analyzing the problem and get solutions into gear (see ch. 7). Maybe we can make the fight against climate change sexy and cool (see ch. 3)? I'm going out on a limb with this one, but if sex can sell almost anything, then maybe it can even sell solutions to climate change.

But there's still always denial. I don't know about you, but I could use a cheap instant vacation, and denial delivers. Unfortunately, for all its supposed health benefits, denial doesn't help us prepare for the insidious risk of sex and a crazy little thing called love, coming up next.

CHAPTER 3

SUCKERS FOR A PRETTY FACE

How Sex and Love Lure Us to Let Down Our Guard

BECAUSE I COULD

I'm sure he knew he was in trouble, but still on camera he managed to look indignant when asked the question. "I did not have sexual relations with that woman," Bill Clinton proclaimed, jabbing his finger at the camera. When the story about Bill Clinton and Monica Lewinsky hit the news, I couldn't believe it. It wasn't that I was swayed by Clinton's denial. I just couldn't see how it was even possible for the president to have the time for such a thing, much less to do it in the Oval Office. I'm not the president, but I barely have time to shave in the morning. I must not understand how government works.

However, I was not the only one who could not believe it. Even Clinton's own cabinet was surprised. "He's got all these enemies who are out to get him," Robert Reich, Clinton's secretary of labor, said in a recent film as he recalled the scandal. "He wouldn't be so stupid as to jeopardize his entire presidency. For what?"[1]

Oh, yes he would. The whole thing sounded implausible, but when the story came out—every lurid detail—there was no denying it. The descriptions of presidential genitalia, the blue dress and its

75

presidential stain, were all there as evidence. We do crazy things in pursuit of sexual gratification.

After hearing the news, my next thought was "Why?" Why would he risk so much? The ongoing war between political parties is not new to our time, and Clinton must have known that a scandal like that would divert attention and siphon support from his efforts. Even Newt Gingrich would have to admit (off the record) that Bill Clinton is a smart man; but how could a smart man have done something so dumb?

It might not have made sense, but, nonetheless, he had done it. Clinton's explanation of "Because I could" seems insufficient, though. There's far more to the story, because he's not the only one. The Hall of Shame is painfully long: Anthony Weiner and his unfortunate photo tweets; Eliot Spitzer, stopped short in his fight against white-collar crime by high-priced prostitutes; John Edwards, first denying everything, including a child he fathered, and eventually unable to deny anything. There are plenty of Republicans on the top-ten list of sexual scandals, as well, such as Larry Craig, arrested for lewd conduct in an airport restroom; or Governor Mark Sanford of South Carolina, who famously disappeared and said he was hiking the Appalachian Trail but was instead in Argentina with his mistress. Oh, and don't forget Arnold Schwarzenegger, whose indiscretions and fathering of a child with a housekeeper came to light after he left office as the governor of California. Such indiscretions are so numerous that they almost seem to be the rule rather than the exception. And certainly the list is not limited to celebrities and politicians; they're just the ones we hear about. Over and over, around the world and throughout history, sex lures us to dash ourselves onto the rocks, smashing our ship and leaving us stranded, high and dry.

THE PERILS OF ONLINE DATING

Men aren't the only ones to lose their minds in pursuit of sex. Online dating has become one of the most popular places for both sexes to meet people, including sites like AshleyMadison.com that even specialize in married people looking for affairs. One reason online dating is attractive is probably because we can sit at home or in our offices and look over potential candidates for love, sex, or both without exposing ourselves face-to-face. The risk seems low, at first at least. But this gets us only so far. Sooner or later, we've got get out from behind the computer monitor. When we connect with people online and we're ready to actually meet, it's an entirely different story. Online profiles don't reveal the whole story, often telling white lies and sometimes not-so-white lies. We have no idea what kinds of risks we're getting ourselves into.

While researching this problem, I talked with "Sheryl,"[2] who considers herself a student of human nature and works in the health-care industry. She knows all too well the risk of AIDS. She knows better than to engage in unprotected sex with a stranger. But as a recently divorced, middle-aged woman, Sheryl was feeling a little crazy several years ago when she signed up for online dating.[3]

Sheryl is an attractive woman in her fifties, but for years she had felt starved for attention and intimacy with her husband. He had told her right from the start that he had a low sex drive and that it was the reason why he had little interest in sex, despite seeming like a great match for Sheryl in every other way. As it turned out, after seventeen years, Sheryl learned that his sex drive was perfectly healthy—he had been directing it to massage parlors rather than to his wife for all this time, even when she practically begged him for sex. "There are men who would rather be with a stranger. That's why prostitutes are in business," Sheryl said when we talked. "That's why Eliot Spitzer did what he did. He knew it was an issue and took the risk of losing his family anyway."[4]

"It's an extremely vulnerable moment for middle-aged, divorced women," Sheryl said, talking openly about a difficult time in her life, something she shares along with many others. "We do crazy things after a breakup or divorce, in the suburban, white-collar lifestyle. There are a lot of risk takers like this out there, both men and women. Men will risk their families for sexual relationships, and I know most do not want to lose this, but they're willing to risk the one thing they value." And all too often, their fears eventually become reality.

When Sheryl went online, she went to a site that specifically targeted married people looking for sex with the thought that pure sex was what she wanted, without any emotional involvement. "Shortly after separating from my husband six years ago, I was desperate for sexual attention, having been deprived for years emotionally and physically." When she connected with "Jerry" through the site, he seemed like just what she was looking for. At first.

"When I met him, I thought he was a little effeminate, but very handsome and funny and smart. He was an actor by trade, but he had worked very little. I immediately jumped into a sexual relationship with him, and he was a great lover." But something wasn't right.

It seemed like Jerry's acting was not restricted to films and on stage, but that he was acting in the relationship as well. After the initial thrill of jumping into bed with him, Sheryl thought it odd that he was studying books about how to be with a woman. Her instincts told her that she was crazy to be involved with a stranger like this, that despite her initial enthusiasm and thirst for intimacy, there was danger lurking in this relationship she had plunged into. While she thought she had been looking for sex, she realized at some point that she wanted more, and she certainly did not want chaos to be unleashed in her life. Soon she listened to her instinct and Jerry was out the door. But that was not the end of the story.

"After I broke up with him, he turned into a psycho—hateful, vindictive, calling me every name in the book," she said. Jerry became angry and forwarded their sexual e-mails to her ex-husband and to her neighbor, e-mails that said far more than she was comfortable with having revealed in this way. "He did a one-eighty and became abusive and threatening." Sheryl realized that "he was counting on me, in hindsight, because he was desperately broke, and his wife was breaking up their arrangement, whatever that was. He was looking for somebody to take care of him, a middle-aged woman with money who lived in a nice area. He was counting on me, so when I broke it off he panicked and freaked out."

Living in suburbia in Southern California, Sheryl knows plenty of other divorced, middle-aged women. "They are frustrated and haven't been satisfied in many ways. I thought I just wanted sex when I went online, but really I did want an emotional connection. I had sexual encounters a few others [*sic*] times after the crazy time [with Jerry] passed, and I left empty from all of them. That's what motivates women to take risks, to be with men we would never be with otherwise. It's about emotional intimacy."

Reflecting on this time in her life, she said, "I dodged a bullet. I'm one of the least likely to do something stupid. I worked in healthcare and know about the risk of AIDS. I don't have my head in the sand. But when I think back, I wonder, where was my head?"

Having unprotected sex with virtual strangers is always a risk, particularly if we have questions about their sexual histories, and exposing our inner lives to turmoil is a risk as well. But our needs that are buried deep in our minds, swirling around in our psyche, drive us to fill them, even when these needs are in conflict with each other and with common sense, leaving us to struggle, stuck in the middle.

THE SEX DIARIES AND THE PASTOR'S WIFE

As the author of *The Sex Diaries Project*, Arianne Cohen has been collecting the anonymous stories numerous people, from ages eighteen to ninety-four, tell about their sex lives. Since the stories are anonymous, people are open and honest about what they do, far more so than the faces they present to the public. We might lie to impress others, or we might hide the truth from others, but we're more likely to be truthful when people won't know who we are. I've got a feeling that a lot of people find it a relief to have a place where they can be honest without worrying what anyone will think. Arianne Cohen has probably read just about every crazy thing that people do in the name of sex, often putting everything they value on the line in pursuit of it.

"Lots of people do risky things," Cohen said, summing up her impressions after receiving 2,500 stories or so submitted to the project since 2007.[5] But beneath it all, there's a method to their sexual madness. "They're trying to get their needs met, their emotional, sexual, and life needs. Sex is a big way of dealing with emotional as well as sexual needs," said Cohen.

One story in the project was from a British man who was in a four-year relationship with a woman while seeing prostitutes the whole time. Some relationships allow outside sexual partners, but his did not and he knew it. "The girlfriend had no idea whatsoever," Arianne mentioned in our discussion. "And obviously she would not be okay with this. But what's interesting to me is what makes it risky:

not the behavior itself (seeing others), but that the behavior is not agreed upon with his primary partner."

Knowing this situation was not okay and that it might blow up in his face, why would he pursue it? "He was getting an emotional release out of it. He also thought that he was expressing sexual proclivities that he couldn't express with his girlfriend—though the reality is that he never asked her."[6]

Another reason for risky behavior is a conflict between a person's needs and what is deemed acceptable—the most common example being people who feel a need for multiple sexual partners. "Our culture lets us know that consensual, monogamous relationships are okay; it is less culturally acceptable to be dating multiple people, or to be in a polyamorous relationship," Cohen says. "We have this idea that people are either in a relationship or dating to be in one, but there is a wealth of ways to live your life. There is no limit to the risky behavior that people will take on, but often [they are] not in healthy relationships to begin with, not communicating, not in touch with their needs," she adds.

Talking with Arianne, I wonder why someone like the British man gets into relationships if, right from the start, he is putting it at risk by going outside the relationship with someone else, jeopardizing his emotional needs for his sexual needs. If he understood himself—if we all understood who we are, what we need, and how to get it—he probably would not do things like this. But we don't understand ourselves, much less other people, for the most part.

"People like him do it because they're not in touch with what their needs are and [they] don't understand themselves," Arianne said. "If you did understand your needs, can express them and act on them, you'd probably find a better way to do it, a healthier way to do it. You would find a healthy, consensual relationship; but we don't."

As the pastor's wife in a small Southern town, Yolanda King Stephen thought she knew her husband pretty well.[7] She learned how wrong she was when a young man, a member of the congregation, came to the house. "It was revealed that he [her husband] was sleeping with young men and some women in the church. A young man he was sleeping with was a family friend," Yolanda said. When the news came out, it shook them all, and it shook them hard.

"In the black church community, homosexuality is a shunned subject," said Stephen, reflecting. Homosexuality is not broadly accepted in many groups, with some groups even more resistant to the topic than others. The risk of going against these social rules can be great in the minds of those who find themselves on the wrong side of the rule. "We don't talk about it and, especially in the South, it is a stigma that causes many to hide their sexuality. He was a pastor with a rising congregation, a trusted member of the community, and he was loved by many." This was too far from the socially acceptable path, and such a revelation was more than the pastor could bear. Soon after the secret was revealed he committed suicide, leaving his wife and two young children behind, struggling to cope. His wife is somewhat philosophical, looking back, but she also still feels a deep pain. To her, as hard as this was, it was something they could have worked through; but for the pastor, it wasn't. He took a risk and broke the rules of society and, ultimately, forced himself to pay the highest price for taking a risk like this.

FISH DO IT, BIRDS DO IT

In looking at risk and how we handle it, you might think we humans are rational thinkers, harnessing the power of our big, beefy intellects to mentally calculate the benefits and risks of decisions before leaping into action. Sometimes we actually do this, or try to, at least. We do have the power to analyze a situation, doing a bit of quick mental calculus, carefully adding up and weighing the odds before taking the action necessary to minimize risks. All of this sounds great. But for the most part, this way of looking at things seems to be just plain wrong, not the way things really work inside our heads at all.

Instead, we make decisions on the fly, moment by moment, usually without skipping a beat. We have a running intuitive feel for what feels good or not, and then we move in that general direction. We leap first and rationalize about risks later, using motivated reasoning, as we saw with climate change denial. Occasionally we stop and really think things through, but not often. And when we do engage in analysis, the body's other systems can override it, sensing that a risk is necessary or provides a benefit great enough to be worthwhile. If the perceived benefit is great enough, the alarm bells get switched off and we go for it. Sex is one system that overrides the risk alarms in a big way.

Part of it is biology. Like all living things, humans have a need and an instinct to reproduce, and at a basic level this boils down to sex. Not that all sex is strictly related to reproduction for humans, but if there is a point to life for organisms on earth, biologically at least, it's about passing on genes to future generations. In evolution, genes that ensure the production of offspring and their survival are those that are carried forward in future generations. Anything that supports reproduction will be heavily selected for, which might explain why the sexual drive is strong enough to drive us beyond all reason. The connection between sex and reproduction is not always as close as it once was, but the reason that sex slams our pleasure centers so hard is that evolution has made it that way. It's not the devil that made you do it—evolution did.

In nature, animals respond to stimuli sometimes in an exaggerated way, particularly stimuli related to sex and reproduction. Moths will fly miles in pursuit of a mate, following the chemical scent of a few molecules of a chemical, called a *pheromone*, that the object of their affections releases to the wind. The power of this scent signal can be used in insect traps, luring moths to their deaths as they blindly and instinctively pursue sex all the way into a dead end.

Another example of how far we animals will go for sex is the stickleback fish. Sticklebacks are small fish, about two to three inches long, with spines sticking out along their backs and with a unique way of finding a mate. Male sticklebacks build a nest in the sand, doing their best to look prosperous and healthy with a nice place to raise the kids, and then they try to attract a female to the nest, where she will lay her eggs to be fertilized by the male. When males build the nest, they not only try to attract females but also fight off other males who might try to get in on the action. And both attracting a female and fighting other males involves the color red.

When looking for a fight to pick, male sticklebacks aren't looking at the whole competitor fish. As a pioneer in the study of supernormal stimuli, Dutch biologist and eventual Nobel Prize winner Niko Tinbergen found that male sticklebacks look specifically at the red other males have on their bellies, aggressively attacking males with red bellies. To see what was involved, Tinbergen started experimenting by putting dummy fish into a tank to act as competitor fish instead of real fish. He found the same thing happening: males would still attack the red spot, attacking the dummy fish even more than the real male if it was red enough.[8] The redder the belly, the stronger the response, no matter how the rest of the fish looked. Males will even attack a circular shape on a stick if it's got enough red painted on its bottom.

Birds also respond strongly to specific clues in their quest for sex, including the brilliant birds-of-paradise species in New Guinea. Male birds of paradise have highly elaborate plumage and mating rituals, involving bobbing and weaving, while females are much more subdued in their appearance. When male birds must display certain colors or behaviors to attract a mate, the stronger the stimuli of the male, the more likely the female is to go for it. While Darwin studied finches in the Galapagos, his competitor Alfred Wallace was lying in the jungles of New Guinea, deep in a malarial fever, arriving at similar conclusions by studying birds of paradise.[9]

Sex has a variety of benefits, starting with genetic diversity as well as the more obvious benefits we associate with it. Having sex allows plants and animals to mix up genes from males and females into new combinations with each new generation, a process that evolution has found to be so advantageous that it's almost everywhere in the natural world, a preferred way to reproduce even if it takes more energy than sprouting off clones of yourself like spider plants. Some organisms, like water fleas, can reproduce asexually whenever they want, without sex, particularly when the environment is rich with nutrients and producing lots of offspring quickly is the key to reproductive success and evolutionary fitness. But the water fleas then switch back to sex when the environment gets harsher and their pond is in danger of drying out, using the genetic diversity created by sex to help the population survive hard times.[10]

Sex can help a species survive, but sex can be risky too. Males with bright colors that attract females can also catch the eye of less desirable neighbors like predators, which is what happens for unfortunate male guppies. Male guppies with orange dots are preferred by females and are also more robust then their less-orange friends in many ways, performing better at finding food, moving around the tank where they are being studied, and acting boldly toward predators placed in the tank. So it might make sense for females to view these brighter males as a good bet for reproduction, with good genes to offer her kids. Fixating on the size of a male's orange dots, a female's instincts tell her that orange dots make the male. Orange turns her on. And she's not the only one. As it happens, a common predator of guppies is cichlids, and cichlids love to gobble up orange males.[11] Having orange dots is risky for males, but it goes with the territory if you're a male guppy and want guppy sex. What's a guy to do?

SEX SELLS

People may not be seeking orange dots when it comes to looking for partners, but we are certainly not immune to stimuli related to sex, seeking to get our needs met one way or another. You see sex in marketing everywhere because it works so well at catching our eye. What does the model draped on top of a car in an ad have to do with automobile performance? Sex sells. And sells, and sells.

While people around the world may not agree on many things, the male view of what makes women beautiful is remarkably consistent in some ways. When shown a range of photos of women, males around the world consistently rank a waist-to-hip ratio of 0.7 as the most attractive, which is in the range of Elizabeth Taylor in her youth.[12] The female face that is the most average is the most attractive to men, a composite image of many faces being judged as more beautiful than any individual face. Symmetrical features are also considered highly beautiful, presumably demonstrating that an individual has suffered fewer physical insults and might therefore possess a desirable collection of genes.[13] The universality of these traits suggests that nature is at work, ensuring we're sufficiently blinded by rewards to ignore the risks.

Besides a pretty face, the secondary sexual characteristics of women, such as breasts, hair, and skin, also drive men over the edge. According to the nubility hypothesis, there's a reason why men find these traits attractive: these traits are a way of communicating fertility to men.[14] Studies have looked at the correlation of reproductive hormones on these secondary sexual characteristics and found that the traits we think of as attractive tend to be associated with fertility in women. Beyond our physical features, humans have a fairly unique sex life compared to most animals; sexually mature humans are more or less up for it all the time, while most animals are interested in sex only in specific periods associated with fertility and reproduction. This no doubt helps to keep humans constantly on the lookout for trouble.

In humans, and in other animals, when males and females are on the lookout for certain traits in the opposite sex, it leads to its own form of natural selection, called *sexual selection*, as we saw with Wallace and the birds of paradise. If female peacocks, for example, are on the lookout for males with brilliant feathers, and males with brilliant feathers have more kids, then future generations of males will have even more brilliant feathers. Humans might not be looking for orange spots or bright feathers, but we do seem to have our own version of sexual selection, looking for our own uniquely human stimuli.

With males on the lookout for female traits, exaggerated female features can produce an even stronger response than the real or natural thing, even in the absence of real sex. Breast implants, cosmetics, and pornography seem like examples of supernormal stimuli for men, providing a stronger response than reality alone can provide, designed to hijack natural impulses and produce an irresistible response.[15] Think Jessica Rabbit, Roger Rabbit's cartoonishly exaggerated wife in the 1988 movie *Who Framed Roger Rabbit*. The stronger the stimuli, the more likely men are to lose their cool, eyes bugging out, and take greater risks.

Women have stimuli that they respond to as well, though they're not exactly the same ones as men. Some of the physical traits women respond to are similar to those that work for men, including symmetry in appearance, and women also favor men who are taller, within certain limits.[16] Women in many cultures tend to marry men who are somewhat older than themselves, perhaps because this allows for greater economic status of the men and provision for potential children. Women with more symmetrical male partners tended to have more orgasms than women with less symmetrical partners.[17] Having more orgasms in turn seems to speed sperm toward their destination (to fertilize eggs), and it also seems to solidify the bond between partners, increasing overall reproductive success.

Some of the stimuli to which women respond are not found in men but in themselves, such as the pervasive images of models and celebrities in magazines. Magazines like *Cosmopolitan* always feature pictures of beautiful women, but *Cosmo* is not targeting men. It's targeting women, and the way in which women compare themselves to others, as described by Deirdre Barrett of the Harvard Medical School.[18] Like pornography, *Cosmo* did not invent the motivations that make these stimuli attractive, although it exploits this drive as much as possible. The magazine plays off of our inner needs and drives, taking them to greater extremes. "'*Cosmo*' models look suspiciously like male magazine centerfolds," writes Barrett.[19] "Here, however, they are icons not for anonymous casual sex, but instead for becoming the irresistible date and mate choice," she adds.

"Anything that sells spectacularly well is probably some type of supernormal stimulus," Barrett continues. Like pornography, we're drawn to *Cosmo* and, say, cheeseburgers for similar reasons: because of the deep instincts they trigger, hitting buttons that can be hard for us to ignore.

Note to Self: invent cheeseburger pornography. Or maybe pornographic cheeseburgers.

THAT'S WHAT YOU GET WHEN YOU FALL IN LOVE

Having sex with someone, or proposing to, comes with a variety of risks for humans, as it does in animals. There's the risk of rejection, the social risk, and there's also the risk of getting a sexually transmitted disease, like AIDS, which is transmitted by the human immunodeficiency virus (HIV). While antiviral drugs have kept HIV under control in many population groups, in some African countries, as many as 25 percent of the population ages fifteen to forty-nine are infected.[20] It is important to note that the United States and other developed countries are not immune, with over a million people in the United States living with HIV.[21]

The development of drugs to treat HIV revolutionized the treatment of AIDS. Whereas there was nothing that could stop the progression of the disease when it first emerged, people now receiving antiviral drugs can live many years looking and feeling healthy, with viral levels kept in check by the drugs. The drug treatments can be complex, though, with multiple drugs involved, each one with its own schedule; and the virus doesn't go away entirely even after years of treatment. Whereas a flu virus gets cleared from the body once the infection is over, HIV lurks in cells practically forever. If the drugs are stopped or the schedule is not followed, then the virus can emerge and start replicating again, often in a new form that no longer responds to the old drugs. The threat never goes away for those who are infected or for those with whom they have sex.

When HIV-positive people feel healthy, they might forget the risk or choose to ignore it. To look at the increasing rate of new infections in the homosexual population, the DNA of viruses in newly infected patients was examined by Prof. Zehava Grossman of Tel Aviv University's School of Public Health. These patients had never been given antiviral drugs, but the virus they carried had already been exposed to the drugs. They were infected by patients who had HIV, knew it, were receiving antiviral drugs, and then apparently had risky sex with unprotected partners. They may have believed that the treatment with the drugs had reduced or eliminated the risk that they would infect others, but it's also possible that the prospect of having sex overrode their concerns about the risk.[22]

Psychology researchers Hart Blanton and Meg Gerrard found that sexual attraction bypasses the mental safeguards that block risky sex, leading to a sudden underestimate for the risk of sexually transmitted disease.[23] Heterosexual undergraduate men at Iowa State University were first asked if they would have sex with a woman before marriage (not a specific woman, just the general question). Only three said no and were thus excluded from the study. The forty remaining students were then given information about several different hypothetical women, such as the number of sexual partners they had had and whether or not they used condoms. Then the men were asked to judge the risks these women would present for the contraction of sexually transmitted diseases, including HIV. Later, the men were shown pictures of women along with other information. The pictures were selected to be either HSA (high in sex appeal) or LSA (low in sex appeal). One can imagine the criteria for the two groups—an additional group of college males were responsible for these rankings.

In this study, the individuals were not specifically identified and were not someone those taking the test knew. There was no expectation that they would actually have sex or anything like it with the women pictured, and there was no risk of actually contracting an STD. Despite this, the attractive HSA women were consistently judged by the men to be lower risk. The men's ability to judge risk was altered by the attractiveness of the women in the photos, causing them to let down their guard.

This might not sound earth-shattering, but it has significant implications. If a picture alone is enough to alter how we see risk, imagine how our risk perceptions are overwhelmed in the heat of the moment. It's not enough to tell men about the statistics for STDs and AIDS if at the crucial moment all rationality gets thrown out the window. When people are at the bar, talking and having a few drinks with an attractive, actual person, these influences could easily overwhelm anyone's ability to carefully weigh the risks involved in a sexual encounter. "It is our hypothesis that behavioral motivations often undermine a rational risk assessment, such that people construct vulnerability perceptions that will justify the desired response to a situation," wrote Blanton and Gerrard, reporting the results from their study. My thoughts exactly. It's easy to ignore a risk if we want something badly enough. And we do.

Sex opens us up not only to the risk of sexually transmitted viruses but also to computer viruses. Most people are aware of the danger of computer hackers today, and millions use some form of antiviral software to protect themselves. They would not typically go clicking onto random links, and are familiar with the problems associated with wire transfers to associates of Nigerian dictators. But when pornography is involved, all sense of danger is thrown out the window; anything goes.

Dr. Gilbert Wondracek from the International Secure System

Lab led a study on the dangers of Internet pornography, attempting to see if it is truly as dangerous as is believed.[24] His team reported that out of 35,000 domains and 269,000 websites devoted to pornography, about 3 percent held dangers like viruses. To see how vulnerable site visitors on the net are, the team went on to build its own porn site and buy visitors, hiring other sites to redirect visitors using hidden tricks that may open up vulnerabilities to users. In their "pornpreneur" study, the team found that out of 49,000 site visitors they bought, 20,000 were vulnerable and a thousand were already infected after failing to engage in safe cybersex. (The researchers weren't themselves planting viruses, and they gave away any money they might have made as a result of their pornpreneurship.) Three percent is far from 100 percent of porn websites, and planting viruses is not how most porn-website operators usually make their money; but a small percentage of sites lurking out there do use pornography specifically as a lure to attract visitors and then plant viruses in their computers like fishhooks.

WHAT'S LOVE GOT TO DO WITH IT?

Of course, sex is often associated with its fair cousin, love, which comes in several flavors. There's romantic love, long-term-commitment love, and there's love lost. Love is a risk, and judging by the wreckage of its aftermath, it's a big one; the scars from the vulnerability of emotional intimacy can run deep. People risk a lot for sex, but seldom do they launch a war over it. Love is another story, as the Trojans found out when the Greeks launched the Trojan War over Helen of Troy. People do crazy things for the sake of love and even crazier things when they lose it.

In her book, *Crazy Little Thing*, Liz Langley lays out some of the crazier stories of love won and lost.[25] For example, there's the story of Burt and Linda in New York, in which Burt was convicted of assaulting Linda when she had lye thrown in her face, scarring and nearly blinding her. Fourteen years later, when Burt got out of prison, Linda married him. There's the story of Arthur and Lynnette, who fell in love in their quest for the reptile aliens they were convinced lived in the Mount Shasta area, melding his Asperger's with her spiritualism. There's cousin love, necrophilia, and countless other permutations on the oldest story. And there's your story, starting right in your own brain.

Love stems from a chemical and neuronal onslaught on your brain. "Guess what? You're not a hopeless feeb because your brain continually returns to the kisses of a certain someone," writes Langley. "You're not a fool or a tool because you've realized that Copernicus was wrong and it is your beloved around which the universe revolves. When love happens there are chemical releases going on that can turn you from Ren into Stimpy. It's up to you to find a balance in between."[26]

By looking at the brain scans of those in love and those whose hearts have been broken, it seems that the same brain regions involved in drug addiction are also stimulated by love. Many of the same hormones that turn on when we're romantically involved ramp up even more when we break up, as described by Langley: "So rather than back off and give us a break, the chemicals of love increase when we are rejected, causing us not to say, 'Oh well, easy come, easy go' and move along, but to hunger for our love object even more. It's like putting a nice warm heating pad on a blistering sunburn. Or like having your soul pulled out through your pores."[27]

Ouch. Definitely not good, and yet we can't get enough of the stuff. You touch a hot stove only once before you learn to stay away;

but with love, we keep coming back for more. Those are some potent neurotransmitters. I guess they have to be or we wouldn't go for the whole "crazy" thing.

James Coan at the University of Virginia measured the physical impact of loving relationships on women. The women were given an electric shock, and their response to the shock was measured, including the pain they felt, their anxiety, and how the shock was perceived in the brain. Then the women repeated the test, doing everything the same, except this time they did it while holding the hand of their romantic partner. Women in a healthy and loving relationship felt less pain and anxiety when holding the hand of their partner, and you could see this in their brain. They felt protected, and actually were protected, just by knowing someone else was there with them. Women in a less healthy relationship did not feel this protection.[28]

Somehow that really sums it up for me, the image of couples holding hands and feeling better about the world and their lives. That's why we do it. That's what we are after. We long for that sense of protection and togetherness. We long to know that we're not facing the universe alone, that there's someone else there with us, someone who cares about us among the billions of humans and endless cosmos. Sometimes we get what we're looking for and sometimes we don't, but we keep on trying. It's that important to us, making it worth the risk.

CROSSING A BUSY STREET

Individuals vary in many ways in their attitudes toward risk, and one of the factors involved is our sex: men and women, as opposite sexes, often differ in their propensity for risk taking. Of course this is a gross generalization (very gross), but studies suggest there's some truth to it. And the reason for the difference is probably rooted, once again, deep in our messy biology.

Mammals like us reproduce through sex followed by live gestation of babies by females (women). I'm thinking you know this part, so I'll gloss over the details. Suffice it to say that the male contribution may take only a few minutes while the female's contribution will involve at least nine months, and probably far more time once the baby is born. This is not to say that all men will bail out after a few minutes, but it does happen. As a result, it's not an even deal. Again, this is not to say that all men are heels, but there is an inescapable biological difference in our reproductive roles that has often had an impact far beyond reproduction itself.

Because females have a lot more at stake with each baby, they are predicted to be choosy when it comes to picking a mate; meanwhile, males would be expected to compete for mates, taking risks to attract them. The risk taking by males can take the form of fighting over a mate (like bighorn sheep), competing for social status (as occurs with gorillas), or presenting a display that turns females on (as with peacocks). Men don't have brightly colored feathers or horns, so perhaps other methods are required, which might include economic competition. Among human males, economic status and wealth are commonly associated with reproductive success. Sometimes diamonds really are a girl's best friend.

"In recent human ancestral times, men who controlled more resources married younger women, married more women, and produced offspring earlier," wrote Daniel J. Kruger, a researcher at the University of Michigan.[29] "During recent human evolution, males who do not have substantial resources or status may have been unable to establish long-term relationships. Thus, sexual selection helps to explain some sex differences in psychology and behavioral tendencies, including the stronger male tendencies for risk-taking, competitiveness, and sensitivity to hierarchy."[30]

Studies and statistics commonly show that men—particularly young, single men—take greater risks in driving, fighting, experimenting with drugs, having unprotected sex, and handling money. Looking across cultures, across countries, this is true in many different ways. The risk taking by males may aid the pursuit of mating opportunities, and it may also help to explain why men live shorter lives than women on average.

In the United States, a boy born in 2012 has a life expectancy of seventy-six years, while a girl born in the same year has a life expectancy of eighty-one years. Even as life expectancies have gone up over the years, men have always trailed women in almost every country.[31]

The male life expectancy does vary at times, though, growing shorter in conditions that encourage or require greater male risk taking, such as economic challenges. The transition of Eastern European countries to market economies in the 1990s may have opened more economic opportunities, but it also created big challenges and stresses for many, particularly men in areas where they remain the primary income for families. A curious, yet related, observation is that the difference between male and female mortality increased during these economic transitions, perhaps as a result of increased stress and risk taking by men pushed to the edge by hard times.[32] You might predict that the recent economic catastrophe would lead to greater male risk taking and mortality over the long run as well. Only time will tell.

Male risk taking can be seen even in ordinary parts of daily life, like crossing a busy street or catching a bus. A study at the University of Liverpool in the United Kingdom watched a single bus stop on a line headed toward the campus, watching to see how closely students timed their arrivals at the bus stop.[33] The bus would pull into the stop every day twelve minutes before its scheduled departure at 9:40 a.m., and it would then wait. If a student arrived too early, they would lose time while they sat and waited for the bus to arrive. If they arrived too late, they would risk missing the bus and be left standing in the cold even longer and eventually would risk missing class.

There is in this situation an optimal time to wait and minimize the risk of missing the bus without wasting too much time at the bus stop. Sometimes the bus leaves a few minutes before the scheduled time, making the least risky time to arrive at the bus stop about five minutes before the bus leaves. When women are considered in the experiment's data, looking just at those who walk to the bus stop alone, they instinctively gauge their arrivals to the optimum, about five minutes beforehand. Single men, however, arrive later, taking a greater risk.[34]

Missing the bus might not seem like a very big risk, and it's probably not life changing. It also does not necessarily reveal why these men would risk missing the bus. In another study, however, the same researchers from Liverpool examined how men and women would cross a busy street. Researchers observed a busy street on campus in Liverpool and judged the traffic when men or women pedestrians approached. The researchers then broke down the amount of traffic present as low or no risk, or as risky when cars were passing through the pedestrian crossing.

When the data were examined, men at the crossing were significantly more likely than women to cross at risky moments when cars were passing in the street. But this behavior was not seen all the time. Most people avoid risky crossings when spectators are present. But young men in this UK study were particularly likely to make risky crossings when females were present, as if crossing the road and taking risks when cars were passing was their way of showing off for the opposite sex.[35] No mention is made in this study if this tactic was successful for the males.

In another study, young male skateboarders took more risks and more falls when observed by an attractive female researcher. The higher the skateboarders' testosterone levels, the greater the risks they took. "Our results suggest that displays of physical risk taking might best be understood as hormonally fueled advertisements of health and vigor aimed at potential mates, and signals of strength, fitness, and daring intended to intimidate potential rivals," said Prof. William von Hippel of the University of Queensland.[36]

Testosterone levels are associated with other types of risk taking as well. John Coates and Joe Herbert, researchers at the University of Cambridge, measured testosterone levels in traders on a trading floor in the city of London.[37] Keeping it real, male traders were examined while they were on the job making real trades, with researchers sampling their saliva during the course of the day and then correlating this with how their trades were going. The sizes of trades ranged from £100,000 to £500,000,000—some good-sized wads of money, large enough to make it real for pretty much anyone. Good days for traders did indeed see their testosterone levels going up, while bad days saw it fall. Testosterone levels have also been observed to rise in the blood of winning male athletes, boosting their performance the next time around and stimulating further risk taking.[38] Testosterone in lab studies has several impacts in dealing with risk, including being fearless, but that's not always a good thing. One would be tempted to suggest that traders take anabolic steroids to boost their performance, except that these drugs are associated with some unfortunate side effects, including taking unnecessary risks (i.e., stupid risks), impulsivity, and mania—traits that are probably not as useful on the trading floor.

In the end, our need for sex, love, and everything that goes with them makes us vulnerable, and there's probably no easy way around it. Odds are, we probably won't stop pursuing these things any time soon. While letting down our guard exposes us to risk, it is also an essential part of connecting with the world around us, including the connections we make with each other. The trick is to know when it's okay to let down our guard and when it isn't, or when we need to raise the alarm for rare risks we often miss, like those discussed in the next chapter.

CHAPTER 4

WHY WE BOTCH RARE RISKS

The Strange Story of "The Big One," 9/11, Katrina, and Deepwater Horizon

WAITING FOR THE BIG ONE

I've lived in California my whole life, so you'd think I'd be used to earthquakes by now. But I'm not. Not even close. They still knock me for a loop every time. I was in San Diego when the Northridge earthquake hit in the early morning in 1994, and even though it wasn't "The Big One" and we were far from the epicenter, it was big enough for me. Parking structures and freeways collapsed around nearby Los Angeles while my wife and I ran out of the house, in the dark, to sit in our car. We sat, breathless, listening to the radio. How big was it? How far away was it? Eventually our heart rates slowed and we headed back indoors to bed. Then an aftershock arrived, and we gave up on sleeping for the next few days.

My wife and I made a list of all the things we needed for the next quake: enough food and water for three days, batteries, flashlights in case the next quake strikes during the night, and a number of other items we might end up depending on in order to survive until help reaches us. For weeks after the Northridge earthquake, we thought another quake was arriving every time a large truck drove by. We were determined to be more prepared for the next time an earthquake hit. Then a funny thing happened: our fear faded away so quickly that I never bought a single item on our list. Today we're just as unprepared for the next quake as we were for the last one. But, as it turns out, we're not alone because, by our very nature, human beings are predisposed to failing when it comes to evaluating rare risks like earthquakes and hurricanes.

Through the millennia, humanity has seen a great variety of risks, some of which happen all the time, while others are seldom encountered. With frequent risks, we learn from our mistakes, navigating past hazards based on countless days of accumulated experience. When we frequently meet a hazard and subsequently get burned, cut, or otherwise injured, we quickly learn to stay away. Our biology sees to that, stamping memories of painful events clearly in our brain.

Rare risks like earthquakes are another story. Most of us will seldom experience risks like a large earthquake, so when they do happen, we don't have the benefit of our experience from which we can draw. In fact, rather than preparing us for rare risks, going so long without an earthquake or other rare hazard teaches us to believe that these events will never happen, at least not to us. This assumption works the vast majority of the time . . . until the day inevitably comes when the rare event does finally strike and catches us completely off guard. It happened with the HMS *Titanic*, the terrorist attacks of 9/11, earthquakes, hurricanes, and floods. It happens with flu pandemics, wildfires, and financial meltdowns. Over and over again, we ignore rare risks until it's too late. The pattern is as predictable as the events are not. "Human beings are hard-wired to believe in their heart and soul that disasters don't happen and won't happen to them," said Dennis Mileti, professor emeritus in the Department of Behavioral Science at the University of Colorado–Boulder.[1] And it happened again on the Deepwater Horizon oil platform in the Gulf of Mexico in April 2010.

NOTHING BAD EVER HAPPENS TO ME

On April 2, 2010, President Obama gave a speech in North Carolina in which he proposed expanding access for offshore oil drilling to a vast new expanse of US coastal regions while saying that drilling was not risky. "It turns out, by the way, that oil rigs today generally don't cause spills. They are technologically very advanced," the president said.[2] At the time of his speech, it had been decades since a major offshore oil spill, and he wasn't the only one who felt that the risk had gone away. The oil industry, federal regulators, and many in the public felt the same way, believing it had been so long and things had changed so much that the risk of a blowout with deep-water drilling had vanished. They believed this even as oil exploration pushed into the ocean floor a mile or more underwater, under crushing pressures. And as the president spoke, a chain of events was already underway on the Deepwater Horizon drilling platform and in the minds of workers that would lead to the largest offshore oil spill in history.

In the weeks and days leading up to the blowout, the well the Deepwater Horizon was drilling in the Gulf of Mexico ran into difficulty and fell six weeks behind schedule, costing BP (British Petroleum) one-half million dollars each day.[3] Feeling safe still, drillers reduced the number of centralizers (a key safety device that keeps tubes lining the well in place) needed to stabilize the well from the recommended twenty-one to only six. The steps required to form a cement seal in the well were shortened or skipped, and the usual test on the cement was eliminated. Fire and gas alarms were intentionally disabled to avoid waking anyone up at night due to false alarms. Step by step, they ratcheted up the risk higher and higher, and still they pressed forward. On the platform the day of the blowout, tests and sensors showed dangerous pressure levels rising in the well, but the idea of a blowout was unthinkable to workers. That is, until natural gas bursting up through the well exploded, claimed eleven lives, consumed the Deepwater Horizon rig, triggered the massive spill in the Gulf of Mexico, and eventually cast a long shadow over BP.

Looking back, it's hard to believe that BP couldn't see the blowout coming—the warning signs were everywhere. At least a few workers on the rig had voiced their concerns, one of them calling it a "nightmare well" in an e-mail between BP colleagues. But earlier in April, before the spill, the world was a different place. An oil spill this large had not happened since the Ixtoc 1 spill in the Mexican side of the Gulf in 1979, when Jimmy Carter was president and the Iran hostage–crisis drama was still unfolding.[4] The Ixtoc spill in the Bay of Campeche spewed oil for almost a year and decimated fishing and ecosystems in the area for years after, but so much time had passed since then that few people remember it. The oil has lasted longer in the mud of the region than the memory stayed in our minds. It was easy in early April 2010 to think that the risk of drilling had disappeared; after all, the belief that oil rigs don't blowout, explode, and cause massive oil spills had been right for thirty years.

Because we are built to learn from the immediate past and because so many years had passed without a catastrophic blowout, the drillers learned the wrong lesson about the potential risks they faced. They learned to believe that the risk had gone away or at least decreased so much that they could relax safety measures and get away with it. If we experienced earthquakes or oil spills every day, we would learn right from the start that the earth is a dangerous and uncertain place, and we would always be prepared for its surprises. Instead, our experience teaches us the opposite. Even if we know intellectually that there's a risk, even a risk that could cost us our lives, thousands of years of evolution conspire to convince us otherwise. We know consciously that it's there, that it could happen; but at a very deep level, we simply don't believe that it will.

UNPREPARED FOR THE WORST

The way we learn from immediate risks has served humans well in the past. From the origin of humanity up until the advent of agriculture five thousand years ago, humans were hunters and gatherers living mobile lives. People survived by fighting or running away, a strategy that worked well at the time against threats like predators. "The way people evolved when we were nomadic was by being short sighted and reactive," Robert Meyer, codirector of the Risk Management and Decision Processes Center at the University of Pennsylvania Wharton School, said in an interview.[5] "If you were somewhere where it flooded, then you moved. It was a good survival instinct. People have an instinct for learning that is very adaptive for repeat behaviors, for which you get positive reinforcements," said Meyer. Life was short and dangerous, and you adapted moment by moment based on what you learned from experience. If you were a nomad and found yourself in peril, running away was usually a good solution.

However, what worked millions of years ago for short-term risks doesn't work as well today. We live in a different world now, with most of humanity living in cities and vast, interconnected societies. We have homes, jobs, electrical grids, freeways, and skyscrapers that are stuck in one place, which makes us vulnerable to natural disasters, such as earthquakes, hurricanes, and wildfires. We cannot wait for disasters to strike before we do something about it—it's far too late at that point. Living in large, crowded cities, we need to plan for rare disasters far ahead of time if we are to successfully deal with them; but our biology is still stuck in the old reactive mode of our ancestors, expecting the best and unprepared for the worst.

"For rare events, we get positive reinforcement for unsafe behavior rather than being safe," said Meyer. If we spend money preparing for a disaster by stocking up on food, water, and batteries, and the disaster never comes, then our instinct is to feel like we wasted our time and money. The same goes for expensive flood-prevention strategies, like upgrading levees and building seawalls. Public officials are prone to the same perspective, avoiding costly disaster preparation in their terms of office with the hope and belief that the disaster won't happen on their watch. When we assume that the world is safe and we'll be okay without doing any disaster preparation, we'll be right almost every day. There are far more days that hurricanes or earthquakes don't happen to us than days that they do. The lesson we learn instinctively is that we are rewarded for failing to prepare.

LEARNING THE WRONG LESSON

Part of BP's problem was that it had been through major safety problems in recent years that could have provided valuable learning experiences, but it had learned the wrong lesson. In 2005, BP had a near miss at an offshore platform and a lethal explosion in its Texas City, Texas, refinery. British Petroleum itself noted: "Our [BP's] own internal report found serious problems with the safety culture at Texas City."[6] The company was fined in 2005 by OSHA (the Occupational Safety and Health Administration) for $21 million as a result of safety violations and again in 2009 for $87 million,[7] not to mention a $50 million fine by the EPA (Environmental Protection

Agency). All of this might sound like a lot of money, but BP earned profits of $17 billion even in the down year of 2007.[8] The potential liability for offshore oil spills was capped by Congress at $75 million, further easing concerns that might have moved BP executives and managers to truly change their ways. Having survived these near misses practically unscathed, things quickly got back to normal at BP, and the pressure for short-term results continued.

The failure of safety measures that lead to the Gulf oil spill might have been a calculated move on the part of some at BP (and this will no doubt be argued at trial). They might have carefully weighed the probable outcomes in a cold, dispassionate manner and decided to roll the dice. It's easy to see all of BP as a cold and calculating corporation, and it has done little to change this perception. Or it may have been that those at BP were just prone to the same human fallibility in seeing and avoiding rare risks that we all have embedded inside us. "They [the executives and managers at BP] learned it was worth the risk and were not thinking through the down side," Robert Meyer said in his interview.[9] Even with teams of risk experts, who were paid big bucks to steer the company clear, BP still got caught with its pants down.

We should keep in mind, however, that BP is not the only one. We learn the same lesson in our everyday lives. Whether or not we wear seat belts, bike helmets, or condoms probably won't affect us today, although the consequences can be catastrophic in the long run. But we don't have a bike accident every time we get on a bike, and we don't immediately contract an STD every time we have unprotected sex. "People who reflect on this event will experience a false sense of security and begin to believe that they are less vulnerable to the threats," write Barbara Kahn and Mary Luce at the Wharton Risk Management and Decision Processes Center.[10] The longer we go without a car accident, bike accident, or STD, the more likely it is that we relax and proceed without safety measures, putting our lives on the line based on a faulty instinct.

KEEPING UP WITH THE JONESES

While we seldom experience earthquakes, they're not all that uncommon when you look at the planet as a whole. Large earthquakes over a magnitude of seven on the Richter scale happen on average fifteen to eighteen times a year somewhere in the world. Even if we live in an active earthquake region we probably will not experience more than a few of these in our lives; but, given enough time, they are a virtual certainty in an active region, like California, which has a 99.7 percent probability of a magnitude seven or greater earthquake in the next thirty years.[11] And there's always a chance that "The Big One" will hit: a monster quake of magnitude eight or higher that could smash into Los Angeles or San Francisco to claim thousands of lives.

These facts don't matter much to most of us, though. Talking about probabilities seems to have little impact on earthquake preparation. Most Californians know on some level that they live in earthquake country—we're famous for it. But we still manage to put the thought aside and do little to prepare, with only 12 percent of us buying earthquake insurance[12] and few of us taking steps to protect our homes like securing bookshelves and breakables.

In a quake-computer-simulation game that Robert Meyer helps run at Wharton Risk Management and Decision Processes Center, subjects are given $20,000 of virtual money to prepare a virtual house for an earthquake by buying options like better walls or a stronger chimney, for example. Small groups of players are told an earthquake will come at some point during the game. Players can then choose between spending their money quickly in order to strengthen their homes or leaving their money in the bank to earn interest and thus have more money to spend on strengthening their virtual homes later in the game. If they make their houses strong enough to survive the earthquake when it hits, they win. Still, almost all of the five hundred people who have played the game end up with destroyed houses, even after playing repeatedly.[13] The game is winnable with the right strategy; but players consistently repeat the same mistake, putting off their virtual-home-reinforcement too long. Every single person thinks he or she can beat the odds and come out ahead, and each of them is almost always wrong.

The failure of players to build strong virtual homes may not make sense, but it happens anyway because disaster preparation is not simply a rational calculation based on the odds presented. It is a social behavior. The best way to predict if someone will build a strong home in the quake game is to see what his or her neighbors in the virtual community did. "We started off by thinking that if we had a virtual community with people playing against other real people they could see, and were allowed to talk and communicate, perhaps wisdom would emerge."[14] What happened, though, was the opposite. People, looked, talked, and compared their homes to others and, because the others in the room always had weak homes, players felt comfortable putting off earthquake-proofing their homes as well. What emerged instead of collective wisdom was collective carelessness.

To see if they could disrupt the downward pull of peer pressure in the game, Meyer and his colleagues inserted a virtual player that would go ahead and build a strong house. Rather than follow the example, however, the other players simply ignored the virtual player, deciding unconsciously to follow the majority. That social pressure is so powerful that even when one person in a room is given the right answer on a card telling him the strategy to win, the other players wouldn't follow. "We had a group of seven in a community," said Meyer, describing one experiment, "six guys and one girl, and as it happens the girl had the right answer. They [the other players] knew she had the right answer, and her card told her to build a very strong house, which she did, but still none of the guys followed her. The guys formed their own community and did not want to follow her."[15]

Hurricane evacuations are a social activity as well. When an evacuation warning arrives in a community, people watch their peers and neighbors to see how they respond and then decide what to do. If your friends and neighbors are not packing their cars, you are unlikely to get in your car either. Hurricane Rita developed in the Atlantic Ocean less than a month after Hurricane Katrina devastated New Orleans, and in a matter of days it developed into one of the strongest storms ever recorded in the Atlantic. As Rita entered the Gulf of Mexico and barreled toward Houston, Texas, as a category-five storm, the evacuation order was given. After the flawed evacuation of New Orleans for Katrina, people saw television reports and weather maps of Rita looming in the Gulf, saw their neighbors packing their cars, and decided it was time to get the heck out. The evacuation plans for Houston called for staged movement of 1.5 million people, allowing for waves of motorists to progress through the highway system in an orderly progression away from the coast. Instead, almost three million people tried to evacuate all at the same time, and nobody was going anywhere. People were stuck in traffic for thirteen hours or more, and the evacuation was a bigger disaster than the hurricane, in terms of human lives. About ninety people died on the road as a result of heatstroke, dehydration, and a fire on a bus loaded with elderly evacuees. Meanwhile, Hurricane Rita caused only a few deaths directly.[16] "It was a phenomenon of mass hysteria of evacuation. Because of Katrina, they all went together, a mass movement," said Meyer.[17]

Sometimes, social pressure blows in the opposite direction, as with Galveston, Texas, and Hurricane Ike. As a low-lying island in a hurricane-prone region, Galveston is unusually vulnerable to hurricanes, despite a seventeen-foot solid-granite seawall. The island was devastated by the 1900 Galveston hurricane that submerged the whole island and killed an estimated 8,000 people. For decades, Galveston had been spared from the natural disaster until Hurricane Ike approached in 2008. With Ike predicted to have a storm surge of twenty feet—topping the seawall—the National Weather Service warned 140,000 people in the area of "certain death" if they failed to leave and ordered a mandatory evacuation of 500,000 people.[18] The response this time was nothing like Rita—in fact, it was the opposite. Even with the unusual warning of death, over 200,000 people stayed behind, 40,000 of them in the "certain death" region. "You get the flip side with Ike, when nobody went," said Meyer. "People saw what happened with Rita, checked what their neighbors were doing, and now almost nobody evacuated."[19]

Part of the reason for the low response to the Ike evacuation order lies in the nature of the Galveston community. If Texas is a place apart, then Galveston is even more so; it's an island in more than the geographical sense. Many in Galveston identify more closely with those in their own community than with broader society, and nothing could be a greater source of alienation than the federal government. Those who refused to leave for Hurricane Ike often cited distrust of the government as the reason why they banded together and stayed behind. It was reminiscent of the quake game in which the community joined forces against the well-prepared female player and banished her from their virtual community. Except this time, the government was the outcast and the situation was all too real. And because the damage from Ike turned out to be far less than certain death for most people, the survivors all instinctively learned that they were right not to evacuate. And this only makes it that much harder to get people to evacuate the next time there's a storm.

PREPPING FOR DISASTER

Not everybody ignores rare risks. While the vast majority of us wildly underestimate rare risks, a few wildly *overestimate* them, with survivalists and those called *preppers* devoting a large chunk of their lives to getting ready for the worst-case scenario. Preppers are convinced that these disasters are imminent and, therefore, buy food, ammunition, and anything else they might need to survive events that would force them to be self-reliant. And they are doing it together, with others. While failing to prepare for disaster is a social behavior for most of us, getting ready for disaster is also a social activity for preppers, who use conventions, blogs, chat rooms, magazines, and groups like the American Prepper Network to join together.

Being a prepper is not just a hobby, it is a response to a deeper need. "Being a prepper is an anxiety-reduction approach," said Art Markman, professor in the Psychology Department at University of Texas–Austin.[20] "These people have a massive failure of trust." They have seen failures like the government's disaster response after Hurricane Katrina, and they worry about relying on others for their basic needs. "We prefer to do anxiety-reducing behavior in groups," said Markman. "Community reduces anxiety. We are geared for this—we don't operate well alone, probably for evolutionary reasons. We're not too fearsome alone, but in groups we are."[21]

HECKUVA JOB, BROWNIE

The impact of Hurricane Katrina on New Orleans, Louisiana, in 2005 seemed like the very breakdown of order that preppers fear, and it galvanized many of them into action. As a city that lies mostly below sea level, situated on the Gulf Coast in a hurricane-prone area, New Orleans is particularly vulnerable to hurricanes and dependent on levees, flood gates, and seawalls. When Katrina headed straight for New Orleans, barreling in from hot Gulf waters as a category-five storm, it had been decades since the city had seen a storm like this. The levees failed, 80 percent of the town flooded, and whole neighborhoods were decimated. The levees weren't the only thing that failed in New Orleans. People were left to fend for themselves in much of the town for an extended period of time, including those in the derelict Superdome. Government and individuals alike failed to prepare for a disaster that everyone should have seen coming.

Katrina was treated by some as an unforeseeable event, something nobody could prepare for—but this was not the first hurricane to hit New Orleans. For years before Katrina hit, people in New Orleans were warned of the dangers of catastrophic flooding every time a large hurricane headed their way, and for good reason. In 1965 the Big Easy was devastated by flooding from Hurricane Betsy, with the Ninth Ward twenty feet underwater. A year before Katrina, Hurricane Ivan was headed directly for New Orleans, and an evacuation of the city was attempted. One national newspaper's headline read, "Direct Hit by Ivan in New Orleans Could Mean a Modern Atlantis."[22] Ivan ended up missing New Orleans, and Atlantis was not re-created that time around. However, Ivan did highlight the glaring problem of transportation in an evacuation for the large number of people without cars and the issue of a lack of prepared evacuation centers. Unfortunately, nothing had changed a year later.

When the evacuation order was given in New Orleans as Katrina approached, even those who could leave often chose not to, despite all the warnings and past near misses. These people thought that nothing bad could happen to them—until it did. Rather than motivating a greater response, the near miss with Ivan the year before had reinforced the feeling that preparation was unnecessary. Researchers looking at evacuation responses for Katrina found that even people who had received the warnings did not always evacuate because of what's called a *normalcy bias*. After so many normal days go by, people tend to keep thinking that everything is normal unless they can see a crisis coming with their own eyes. It had been decades since Hurricane Betsy flooded New Orleans, and great faith was placed in the levees and pumps around the city. All the days that had passed without a storm and all the near misses simply reinforced the belief that New Orleans's luck would go on forever.

Another factor that influences our evaluation of risk may be, in part at least, the way we learn about these events, whether they are described to us or we experience them for ourselves. When we read a weather report or hear about people who win the lottery, we are learning from description. In other situations, we learn from experience, from giving things a try and seeing what happens. A research group that included psychology researcher Ralph Hartwig from the University of Basel, Switzerland, put subjects in a simple gambling test to compare these two ways of learning about the world: giving subjects information through description or through experience.[23] The subjects were students in Israel who were given a choice of two gambling options that varied in the frequency of payment. One option frequently paid a small amount of real money (about one dollar), and another option paid more money but did so less frequently (paying about ten dollars in one out of ten turns, for example). One group of students, the description group, read on a computer screen how each of the two buttons from which they could choose would perform before they played. The other group, the experience group, was given a choice of two buttons on a computer screen with which they could experiment; one button on the screen yielded frequent small payouts, and the second button on the screen produced a rarer—but larger—return. For example, if button number one pays one dollar every time, and button number two pays fifteen dollars in one out of ten tries, then, with enough tries, the higher-risk button will pay off more in the long run.

Before playing, subjects experimented with the buttons as much as they wanted in order to learn how the buttons behaved. The two groups faced the same odds, but their responses were dramatically different. The experience group acted as if the rare payoff would happen even less often than it actually did, giving it much less value than math alone would dictate. For those in the other group, who were deciding based on descriptions, the pattern was the opposite. These students were hitting the button with the infrequent payoff far more often. From this, we can gather that experience leads us to overlook the possibility of either positive or negative outcomes if we try something several times and never see a result.

TSUNAMI ALERT

Tsunamis don't happen every day. But when a large tsunami strikes, it can be truly catastrophic, the most lethal of natural disasters. They start with an earthquake on a fault in the ocean floor where an oceanic tectonic plate is slowly sliding, grinding, and sometimes jerking its way beneath a neighboring continental plate. When the plates are moving smoothly, they can move a few centimeters a year. When they stick against each other, they can lock up in a region for hundreds of years or more, building up tension between the plates that is released all at one time in an earthquake. The plates can jerk fifty feet or more when the tension is finally released, and they can move up or down along a plate for hundreds of miles, shoving vast volumes of water out of the way. When this much water is set in motion, it stays in motion, speeding silently across the ocean at hundreds of miles an hour. The water doesn't rise until it nears land, pushing far inland with unstoppable force, sweeping aside everything in its path.

The Indonesian tsunami in 2004 killed over two hundred thousand people in the Indian Ocean region, with a wave that was over a hundred feet high when it hit the island of Sumatra—that's high enough to submerge a ten-story building. A wave this large won't hit a given spot frequently, but over time waves like this repeat themselves quite regularly in some parts of the world where tectonic plates are ramming into each other along the rim of the Pacific Ocean and in the Indian Ocean. The region around the Pacific Ocean and extending to Sumatra and the rest of Indonesia is so tectonically active and suffers earthquakes so frequently that it has come to be known as the Pacific Ring of Fire. The disaster in 2004 was not the first time a tsunami struck this region (and it probably won't be the last), but a tsunami this size does not happen every day, so our experience of tsunamis is almost always limited.

Sometimes people have only moments to act when an earthquake strikes a nearby area. When the 2011 Tohoku earthquake struck Japan in March, those located closest to the epicenter had mere minutes to head to higher ground. Many did, but not everyone. Over sixteen thousand people died that day; some of whom had been warned but decided to stay, thinking they would be safe. Surviving the first wave, others thought they were in the clear, but tsunamis can come in a series of waves, which can, as in this case, last over a period of hours. Still others had survived a previous tsunami in 1960 and thought they would be safe this time. An eighteen-foot-tall seawall had been constructed to protect the town of Minamisanriku—more than enough based on the predictions of the largest expected tsunami. Minamisanriku mayor Jin Sato was there.[24] "Seismologists told us to prepare for a tsunami five and a half to six meters high. But this one was three times that height." The town was obliterated in the tsunami, and any who relied on the seawall did not last long.

Paleoseismologist Kerry Sieh in Singapore has dug into the geological record, tracking tsunamis in the past. By 2003, Seih had uncovered a disturbing pattern. It seemed that giant earthquakes occur in the ocean near Sumatra every two hundred years or so, in pairs about thirty years apart.[25] And the first quake in 2004 was only the first in the pair, leaving the region primed for another in the not-too-distant future: tomorrow, next year, or in a few decades. Or maybe never, so we'd like to believe.

FLIGHT OF THE BLACK SWANS

There was a time when all swans were thought to be white. The very definition of a swan was that it was a large white bird with a slender, graceful neck, and so on. By this definition, a black bird could not be a swan—it was unthinkable. That is, until someone found a black swan and turned this definition and our experience of swans on its head.

Nassim Nicholas Taleb used this analogy to describe rare events and their impacts on us in his book *The Black Swan*.[26] The "Black Swans" Taleb describes are rare, unpredictable events that have big impacts on the world, like the assassination of the archduke Franz Ferdinand that lead to World War I or the precipitous collapse of the Soviet Union. Looking through the centuries at major developments in history, art, or science, Taleb finds these events, such as the terrorist attacks of September 11, 2001, are so rare and unusual, so far beyond the experience of anyone, that they were unthinkable before they happened. No computer model, statistical analysis, or scientific method can tell you that they are coming, although in hindsight we can rationalize how they came about. And looking back through time, these Black Swan events play a huge role in shaping the world in which we live. Statistics might describe the ordinary, day-to-day aspects of life, but Black Swans dominate the larger picture of the world over time.

Taleb writes:

> What we call here a Black Swan (and capitalize it) is an event with the following three attributes. First, it is an outlier, as it lies outside the realm of regular expectations, because nothing in the past can convincingly point to its possibility. Second, it carries an extreme impact. Third, in spite of its outlier status, human nature makes us concoct explanations for its occurrence after the fact, making it explainable and predictable.[27]

In the business world, models of market behavior often assume predictable behavior based on the statistical distribution of the normal curve. Most of us have encountered the normal distribution as the bell curve: shaped like a bell, with a large hump of the most common occurrences around the average, and tails of less likely events on both sides. This works pretty well for things like the height of men or women in a group. Men all cluster in a bell-shaped distribution around the average height of five feet nine inches, with tails sloping off on both sides of the bell, the tails of the curve. Nobody shows up as ten feet tall or zero feet tall—there are no Black Swans when it comes to height.

Markets are not limited in this way. In truth, markets feel no compulsion to limit themselves to our expectations based on the normal distribution or any other model. While this statistical analysis might describe things pretty well in the middle and in the day-to-day ups and downs of the market, it breaks down in the tails of unlikely events spreading out on both sides of typically routine events. By assuming a normal distribution in models, and by ignoring their inability to model extreme outliers like the events that shook the markets in 2007 and 2008, the financial industry and everything it touches (which is pretty much everything) is vulnerable to Black Swans. The models designed to avoid risks cannot capture Black Swans, and so the models eventually fail when the impossible becomes inevitable and the risks arrive all the same.

In many ways this reflects our expectations of the world. Our experience guides our thoughts about the world, based on all that we have seen over the years. Since Black Swan events are so rare, they are not part of our experience, and we are never ready for them. But, nonetheless, they keep on coming. And when they do, they make quite a splash.

WHY IS COKE LIKE EARTHQUAKE PREPARATION?

Studying why we prepare for risks (or why we don't prepare), professor emeritus Dennis Mileti has found that it's more than just a failure to see risks. "If you're talking about the average person, humans don't perceive risk," says Mileti. "Humans think that they are safe. And every day that some high-consequence event does not occur is confirming evidence that they won't worry about things that are low probability."[28]

Mileti and his team talked to people all across the country

in a "Manhattan Project of risk," as Mileti calls it, surveying over three thousand Americans to see what really gets people motivated to prepare for disasters. Funded by the Department of Homeland Security to see what motivates people to prepare for terrorism, the findings of Mileti's study are just as true for other hazards. After sifting through mountains of data and examining every possible variable to explain our behavior, the conclusions from the study were actually quite simple.

"We know conclusively two things and only two things that motivate people to prepare," says Mileti. "The first factor is when people in your personal life share what they have done to prepare."[29] In other words, if your neighbor invites you over to his house and happens to show you how he secured his breakables and strapped his bookshelves to the wall, you will take notice. You're much more likely to go home and secure your own bookshelves. Nobody has to tell us what to do—it's monkey see, monkey do. By seeing what others do, we absorb the behavior and make it our own.

"The second factor is providing dense preparedness information. Ongoing, nonstop relentless information about getting prepared,"[30] says Mileti. Information about disaster preparation finally gets us out of our chairs when it is delivered to us in many ways, often, and for a long period of time. It slowly sinks in when it's on billboards, flyers, television, postcards, Internet advertisements, bus stops, grocery-store bags, coloring books, the morning news, blogs, refrigerator magnets, and social networking. Keep it up long enough and eventually it sinks in and has an impact. "Just the way they sell Coca-Cola," says Mileti. "They still spend millions advertising the product, even after one hundred years. Why? Because if they stop, people stop buying it."[31] Of course, we can't afford to do that for rare disasters and so the result is that people simply don't prepare.

While fear may seem like a good tool to scare people out of their chairs and into hazard preparation, there's just one problem: "It does not work," said Mileti. "It's been known since 1952 that if you tell people how awful something will be, they will turn off. The result can be worse than ever, like trying to scare people into going to the dentist."[32] Most people know they need to take care of their teeth but don't love going to the dentist. If you lay too strongly on the fear of having bad teeth, with talk and pictures of the dire consequences, people tune out the message and turn away, doing even less preparation than before.

While we might underestimate rare risks beforehand, thinking they will never happen, our perception of these risks spikes enormously after they happen. Suddenly, after we experience an earthquake or a hurricane, we become jumpy, seeing risks everywhere. Our feelings about terrorism follow this pattern to a T, particularly after the terrorist attacks of September 11, 2001, that brought down the twin towers of the World Trade Center in New York City and damaged the Pentagon with hijacked commercial airliners.

Before that day, most people felt that the risk of terrorism was low in the United States. The Oklahoma City bombing, the Atlanta Olympics bomb, and the first attack on the World Trade Center put terrorism on the list, but in a Gallup poll conducted just days before 9/11 (September 7–10, 2001), less than 1 percent of Americans felt terrorism was an important national issue.[33] Then, in one day, everything changed. Terrorism leaped to the top of the list in a single bound. In October 2001, a Gallup poll found 46 percent of Americans who participated in the poll saying terrorism was the most important issue the country faced.[34] Everything else suddenly paled in comparison with the images of jets, flames, smoke, and rubble still fresh in our minds.

It's hard to determine the risk of terrorism precisely because it is so rare and sporadic. Worldwide, there were 2,527 terrorism-related deaths throughout the 1990s, according to the CIA,[35] while there were about 560 million deaths for any reason worldwide in the same time period. This makes terrorism the cause of 0.00045 percent of deaths over this period. According to Michael Rothschild, a former business professor at the University of Wisconsin, if terrorists were to pull off a 9/11-scale terrorist act at the rate of one per year, your odds of getting killed in such an attack would still be only one in one hundred thousand[36]—lower than the risk you face as a pedestrian (over four thousand deaths a year in the United States) and much lower than you face driving to work every day (more than forty thousand deaths a year in the United States).

Looking at all possible causes of death for the 310 million Americans living today, about 6,700 or so will die every day, as described Paul Campos, law professor at the University of Colorado.[37] About 50 of these 6,700 will be murdered, 85 will commit suicide, and 120 will die in car accidents. Over 2,000 will die of cardiovascular disease, and 1,500 will die of cancer. Guns will account for about 85 deaths, and work-related injuries for about 15. And terrorism? It doesn't make the list. The number of deaths attributable to terrorism is so close to zero as to be negligible.

There are plenty of reasons why terrorism suddenly seized our attention on 9/11 while crossing the street did not. Terrorism is malicious and out of our control, and the footage of the twin towers from that day includes some of the most dramatic and shocking images of our time. People flung themselves from the burning towers, rubble flew through the streets, and rumors circulated that more airplanes were going to take on more targets. The idea that there are people out there who want so badly to hurt us that they will fly airplanes into skyscrapers and kill themselves in the process is bone-chilling.

While it's not all that surprising that we fear terrorism, it is surprising how many people fear that they will be attacked personally. It's not just that we fear that the country will experience another terrorist event—we are afraid for ourselves. A Fox News poll in mid-October 2001 showed that 46 percent of respondents believed that they might be exposed to bioterrorism agents like anthrax or that a family member might.[38] They took precautions with opening their mail at home, and so many people canceled plans for flying after 9/11 and drove instead that there was a significant increase in highway deaths. The issue of terrorism was not an abstract national issue, it was a threat Americans felt at home and throughout their lives.

Unlike other catastrophic events, the impact of the 9/11 terrorist attacks didn't pass quickly. While our sense of risk always jumps after an event like that, we usually return to normal very quickly as day after day without an attack once again convinces us that we're safe. But a 2006 Gallup poll, five years after 9/11, found that 45 percent of Americans who participated in the poll were still concerned that they would be the target of terrorists, despite no major terrorist events taking place in the United States since 2001 (the shoe bomber and the undies bomber were unsuccessful).[39] And not only did the fear last, but it also extended across the entire country. A 2008 Michigan State study found that the amplified fear of terrorism persists in Michigan, a place never struck by a terrorist attack and not considered a high-profile terrorism target.[40] The fear was so prevalent in Michigan that the state police produced an informational video, called *7 Signs of Terrorism*.[41]

Why has the fear of terrorism persisted when other major events fade so quickly? Remember that Dennis Mileti's research shows there are two conclusive ways to influence how prepared someone will be for a disaster. The first is to look at what the neighbors are doing. The second is to provide ongoing, relentless information about the event. And that's exactly what the media did, day after day and year after year, following 9/11. The footage of the twin towers collapsing has been played innumerable times and the nonstop torrent of information about terrorism from our twenty-four-hour news cycle acted just like Coca-Cola's continuous advertising. Except instead of reminding us to buy Coke, it reminded us to be afraid of terrorism. When we have seen something like this so many times in various media coverage, the anxiety about the risk is kept alive, like blowing on hot coals. Timur Kuran of Duke University calls the media impact an "availability cascade."[42] The more the media talked about this event, the more anxious we became, snowballing into even greater media coverage and still greater fears. The fear takes on a life of its own. The anxiety over terrorism hit some people so hard that a rash of unexplained medical ailments after 9/11 was eventually traced back to high levels of media exposure. People literally worried themselves sick over it.

When a disaster hits you or someone you love in a life-threatening way, it leaves a lasting imprint. Those who lost family members in Hurricane Katrina or on September 11, 2001, will view the world differently for the rest of their lives. And if a disaster is big enough, the world can be changed as well, leading, for example, to improved flood defenses for New Orleans and a slew of new airport security measures. Sometimes a catastrophe is what it takes to shock us into action, taking a threat and transforming it into a game changer, for a time at least. Since Katrina, New Orleans has received the best flood defenses it has ever had, with a $15 billion upgrade to its system sufficient to withstand another storm the size of Katrina, expected once every one hundred years.[43] The town also has a new seawall protecting against storm surges up to twenty-five feet high. It has new pumping stations, floodgates, and miles of improved levees. While efforts to upgrade flood defenses after Hurricane Betsy petered out over the decades as the risk receded in memory, the new defenses since Katrina have been implemented in record time (which is likely the only way they'd have been implemented at all).

There's only one catch: for all they've done, it still might not be enough. As good as the new defenses are, engineers worry they would not withstand a category-five hurricane. While such a storm happens very infrequently and may impact the New Orleans region only once every few centuries, they do happen. Thus, we are still leaving the future open to the risk that New Orleans could flood again, even if we think today it will never happen. And rising sea levels would make this all the more likely.

While our sense of risk spikes after a disaster, it does eventually subside, even if it takes a while. By September 2010, only 1 percent of Americans felt terrorism was the most important problem we face, back to the low levels seen before 9/11, replaced finally by other fears such as the "Great Recession."[44]

Usually the impact of a disaster on our elevated sense of risk is far shorter. "The first nonevent kills it. It goes away quickly," said Robert Meyer, speaking of how long it takes for the sense of heightened risk to disappear after a disaster.[45] Meyer looked at patterns in Internet searches people living along the Florida coast made as hurricanes approached. As a hurricane bore down on a region, more people jumped on their computers and started searching for information about preparation. The more anxious they were, the more likely they were to start searching. But as Meyer sifted through the data about Internet searches, he noticed that residents of one particular county seemed much more worried than others. "We could not figure out why this one county cared a lot," until Meyer and his team saw that it was the county that Hurricane Dennis had impacted a month earlier. And in the same Internet-search data, they found the end point of the reaction as well. Hurricane Ivan had hit a neighboring county a year earlier and caused significant damage, but that county had no apparent surge in searches. People had already forgotten and had gone back to their usual lack of concern. "The worry goes away," said Meyer.[46]

It is possible for us to do a better job of seeing and preparing for these rare risks, but it's not easy because failing to see these risks is in our nature. Even experts who study these risks for a living are not immune from the failure to prepare. When I spoke with him, Dr. Meyer told me of the first year he moved into a house in Miami, Florida. "There was a hurricane threat, so of course I stocked up on water, canned food, and had a whole room full of supplies. Then the season came and went, and another, and eventually my wife said 'What will we do with this water?' So eventually we used it for watering plants."[47]

CHAPTER 5

THE BALANCING ACT

The Give-and-Take Game of Driving and Financial Collapses

MECHANIZED DEATH

I was more than a little nervous when I was learning to drive. In our high-school driver's education class, we spent hours watching movies like *Mechanized Death*, featuring bloody car accidents, designed to scare us teens straight into careful driving. Then, after watching the mechanical mayhem, we got behind the wheel and hit the road. I remember driving with the driver's education instructor sitting next to me in the passenger seat, I think he was also the PE teacher, as I hit the gas and eased out of the parking lot. I had a palpable sense that allowing me or anyone else to hurtle down the road encased in thousands of pounds of metal was sheer lunacy. I was probably right. Most kids couldn't wait to get on the road, license or not; but for me, the first few times behind the wheel were nerve-racking. Terrifying, really. My palms were sweating and my heart was racing as I held the steering wheel straight-armed in front of me at ten and two o'clock. I'm sure my driver's ed teacher felt the same way, cursing under his breath, gripping the arm rest and wishing he'd stuck to teaching PE as I careened around a curve.

My terror soon faded. I quickly got used to relaxing while driving, and was finally able to do the normal level of stupid stuff we do in cars. It's probably a good thing, up to a point. It would be hard to drive much if I remained constantly white-knuckled, arms straight out on the wheel all the time. Losing the fear was necessary if I was ever going to get in the driver's seat on a regular basis.

While most animals each have a very specific environmental niche in which they live, one to which they are tightly adapted through the power of evolution, one of the key traits of humans is our adaptability to live and work in a huge range of conditions and an equally wide range of risks. Humans can get used to almost anything, living everywhere on the planet, from pole to pole and even on board the International Space Station just inches away from the inhospitable vacuum of space. If you move from a New Mexico ranch to a Manhattan flat, at first you might not sleep much with the sirens and constant street noise, but with time you'd learn to live and sleep in this new environment. Likewise, moving from New York City to rural Colorado, the crickets and silence would keep you awake at first, but soon you'd be right at home. For a slightly extreme example, consider that during World War II, people in London slept in subway tunnels for protection as bombs fell on the city. It might not have been the coziest of accommodations, but eventually they learned to manage.

Our inborn ability to adapt to a changing world helps us to survive, but it also leads us to tune out common risks and feel safe, lulling us into carelessness in everyday tasks like driving. Driving is one of the biggest risks millions of us take on intentionally every day, with between forty thousand and thirty thousand Americans dying in their cars each year[1] and over 1.3 million people globally dying each year in car accidents.[2] More people die in cars than by any other accidental death, far more than are killed by homicide, suicide, and terrorism. But nobody has launched a war on cars—it sounds silly for something so ordinary. The very ordinariness of it is part of what makes it so risky.

Driving doesn't feel all that dangerous most of the time because we grow accustomed to ordinary risks like this that we face every day. What once felt risky can soon feel routine. Biologists call this *desensitization*, a way in which something that initially jolts us with a large reaction produces a much smaller reaction later on, after we're exposed to it a few times. For example, if you poke a snail in its eye out on its eyestalk, it feels threatened and pulls its eye back to protect itself from your marauding finger. If you poke its eye repeatedly it soon ignores you, growing desensitized to your annoying pokes. It realizes, in its simple snailish way, after so many repeated pokes that your poking is not a real risk and then goes about its usual snail's-pace business. It learns to ignore you.

We may not have eyestalks, but our responses to seeing the same hazards over and over are similar: we withdraw initially and then learn to ignore the problem and go about our business. In this and other ways, our sense of risk around us is not constant. It varies, depending on what's going on inside us and what's happening around us. Risk is a moving target.

THE CURSE OF INVULNERABILITY

One thing that changes our sense of risk is experience, which usually comes with age. Experience in driving can help desensitize us to its risks—although this does not always make it any safer, particularly for teens. Teens are notorious for their propensity for driving cars into each other, or anything else that gets in their way. Car accidents are the number one cause of death for US teens aged sixteen to nineteen; teens have four times more accidents per mile driven than elderly drivers.[3] One explanation would be that teens lack training in the motor skills involved in driving, or because they lack the experience needed to respond to new situations that emerge while on the road or in the parking lot. Teens do indeed lack experience when they start out on the road, but this isn't the whole problem. The problem with teens comes not just from their driving skills but also from the way many teens perceive themselves. The problem is that they see themselves as invulnerable.

Most of us suffer from an optimism bias—believing that bad things may happen to other people but not to us—but some teens go way beyond optimism, believing that they are invulnerable to everything from car accidents to nasty gossip. It's not enough that these teens know everything, they also believe that they can do anything and not be harmed.

To see exactly how invulnerable teens feel, Prof. Daniel Lapsley at the University of Notre Dame surveyed 350 college undergraduates using the adolescent invulnerability scale (AIS) to look at the students' attitudes.[4] The AIS asks teens to rank statements such as "Nothing seems to bother me," "I could probably drink and drive without getting into an accident," and "Safety rules do not apply to me," from "strongly agree" to "strongly disagree." To look at their propensity for risky behavior, the teens also filled out a questionnaire asking about drug use, alcohol, fighting, vandalism, and reckless driving.

We often wonder what teens are thinking when they do crazy things. It's not that teens don't think, but that they don't think in the same way as children or adults do. In fact, in the invulnerability surveys, some teens revealed that they feel invulnerable to everything from getting hurt by cars to getting hurt emotionally, and these feelings then translate into actions. Teens who reported feeling invulnerable in dangerous situations also reported doing riskier things like driving after drinking. If you feel like Superman, you don't have much of an incentive to play it safe.

One explanation as to why teens might feel invulnerable is that nature makes them this way. Maybe human evolution has given teens this feeling of invulnerability because it is adaptive in some way, helping them to make their way in the world. It's hard to find anything adaptive about drunk driving by teens (or by anyone else), but the world in which teens live has changed from millions of years ago, when booze and cars were not around. Not to say that teens don't have challenges in their lives, but *Homo erectus* teens coming of age a million years ago would have faced far greater risks from predators and starvation than your typical teen faces in modern suburbia. In the risky world of the ancients, being able and willing to do risky things was probably highly adaptive. It might even have its place today but can lead to tragic consequences when the most dangerous thing around is our own behavior.

Laurence Steinberg is a psychologist who studies risk taking in teens, and he has come to think that teens do risky things as a result of their biology and brain development.[5] By imaging a teen brain in an MRI machine, Steinberg has found that the teen brain develops unevenly, with some regions coming online sooner than others.[6] The part of the brain that responds to emotional and social situations is highly developed and active in teens, and, as a result, they often worry a lot about what others think of them.[7] Remembering the social pressure of the teen years, this is not a shock. Development of other brain regions lags, however. The cognitive portions that regulate our behavior, putting a brake on the crazier ideas we come up with, don't fully develop until our midtwenties. The net result? Teen drivers. And being highly socially sensitive, and with few brakes on their behavior, male teens go particularly nuts when you put a few of them together in a car on a Friday night.

Adolescence is a crazy time for humans; but mammalian adolescents of other species also engage in risky business, even if they aren't driving cars. Adolescence is a time when young creatures—whether they are coyotes, mice, or people—develop sexually, become independent of their parents, explore the world around them, and find sexual partners. Adolescent mice are restless, impulsive, novelty seeking, and risk taking.[8] Sound familiar? While most mice don't like going out in the open when it's bright out, afraid of getting snatched by cats, adolescent mice are more likely to take the risk and venture out. Taking risks and doing things that look crazy to grown-ups might make much more sense when viewed in this light. "If it didn't happen, we wouldn't leave home and reproduce," said Steinberg.[9] It's much more difficult to reproduce in your parents' basement than out on your own.[10]

In addition to feeling invulnerable to car crashes, some teens can feel a special protection from emotional harm. Being a teen can be crazy, frustrating, thrilling, and awful, full of immense opportunities and huge barriers as well. The more vulnerable you feel at a time like this, the less likely you might be to take big steps like asking someone out, setting out for a new school, or getting a job. Feeling invulnerable could provide the security needed to take on the challenges and go further in life, even if the invulnerability is not based in reality. In Daniel Lapsley's AIS study, young people who felt emotionally invulnerable actually seemed to do better in life in some ways, being less likely to suffer from depression or low self-esteem. Feeling invulnerable may be a good thing if it protects you against grueling social pressure in high school or from a rough situation at home, though it is not likely to shield you against car accidents.[11]

Knowing that the teen brain gets them in trouble, it would be good to create educational programs that steer them away from the crazier things they might do to hurt themselves. A Toronto study asked high-school students (262 of them) how likely they were to encounter a variety of risks and found that they not only made large underestimates of the risk that they would get in a car accident, but they were also particularly blind to their own role in causing accidents.[12] The teen sense of invulnerability is at play here again, and it seems that educational programs have little impact on this. The mental shield of invulnerability is a strong one, resistant to a variety of assaults. When teens drive, they ascribe risk to external factors, such as automobile and road design, rather than to how they drive the car. They believe that their youth is a huge advantage on the road, providing them with superior reflexes compared with other drivers, and that even if there was an accident, they would not suffer any great injury because doctors would take care of everything.[13] One way or another, the minds of teens cover all the bases, protecting them from every angle, except the risk that comes from themselves.

Putting students in actual car accidents might change their sense of invulnerability, but it may be hard getting permission to run a program like this. An educational program that pierces their veil of invulnerability without causing bodily injury may be preferable.

"They really think that because they're young that they could survive anything, that young people don't die in hospitals. It's only old people [who] die in hospital[s], even if you're in an accident," said Dr. Najma Ahmed, assistant trauma director of St. Michael's Hospital at the University of Toronto.[14]

So, rather than sitting in a classroom, the Toronto researchers took kids to the local intensive-care unit at a hospital to see firsthand other teens who had been in accidents. Seeing what an accident can do to someone—and not just anyone, but someone their own age—finally hit home. After the visit, the teens were asked to assess their risks again, and this time their risk assessments were closer to reality, at least for a time. But it did not last. When the kids were surveyed a week after their visit to see the accident victims, they showed a large increase in their sense of vulnerability compared to before the visit. But when they were surveyed again a month after the visit, the boost in their risk perception had already faded. They had bounced back to their usual invulnerability, with some inner force convincing them once again that nothing bad would happen to them, and that nothing bad *could* happen to them.

Feeling invulnerable may not be realistic and can be dangerous at times, but it can also have benefits in life beyond the teenage years. When successful athletes are examined, one of the traits they possess is their ability to lie convincingly to themselves about their invulnerability. They convince themselves that they will win, and results show that those who do the best job at lying to themselves actually win. People who view the world more realistically, seeing all its hazards and problems, judge their own fates and risks much more pessimistically. The pessimist view of the world may be more realistic, but it can also cripple people's ability to rise above the problems they face.

GIVE AND TAKE

Teens may feel invulnerable, but the feeling passes for most of us as we age and reality teaches us that we are all too vulnerable, leading us to throttle back on the risks we take. But we don't throttle things back completely. You might think that the goal in life, and in your car, is to have zero risk, to create bubbles of absolute safety that follow us everywhere we go, wrapping ourselves in Bubble Wrap or wearing

Kevlar sweats. Sometimes discussion of safety seems to assume that every life must be saved regardless of the cost. In practice, though, things don't work out this way. One problem with seeking a zero-risk life is that it just isn't possible. Instinctively we know that the only way to have no risk is to sit immobile, which would not prove very rewarding (and, furthermore, being sedentary carries its own risks). Every time we engage in various tasks such as getting in a car, we instinctively feel out how much risk is acceptable for that task, given what we're getting out of it. We then let the risk rise up to this point, pushing our risk higher if the reward is great enough. Rather than searching for zero risk, we automatically balance risk and reward using what's commonly called a *risk thermostat* in our mind.[15]

If you're a wild animal, your risk thermostat would kick into action at the local watering hole. You would sense that predators hang out at the watering hole and that approaching the water's edge with your head down is a vulnerable moment. If you're not too thirsty, you'll stay away from the water rather than risk it—the risk is greater than the reward. If you *are* thirsty enough, the reward of getting a drink will push you to literally risk your life. The more vulnerable you feel at the watering hole, the greater your thirst needs to be before you decide that you don't give a damn about lions and go for it.

It's not all that different for people. If you are driving to work and you feel safe, and you've got a strong incentive to go faster (such as being told you'll be fired if you're late again), and you don't feel much risk of punishment, then you might step on the gas until your risk rises to the level you are comfortable with. If it's foggy when you're driving home and you're feeling anxious, and you don't have to rush to be anywhere, you might ease off the gas and slow down, bringing the risk down again to a level that feels okay for that occasion.

We humans are constantly thinking about risk, whether or not we realize it. In fact, as I was casually reading *The Hunger Games*, I couldn't help but notice the role of the risk thermostat and how it played into the story. In the book, twenty-four contestants are thrown into an arena to fight for their lives, battling down to the last boy or girl. But the other contestants are not the only danger these teens face. They're also up against the threats of starvation, freezing to death, and dying of thirst. The organizers of the game create a lake as a tempting source of water, but the contestants know that going to the lake makes them vulnerable to attack. Katniss Everdeen, the sixteen-year-old heroine of the story, is constantly torn between her need for water and food and her need to survive from the others who are hunting her. She is also tempted by berries when she is in great need of food, then she remembers that they might be poisonous.[16] The story of the risk thermostat plays a big part in *The Hunger Games* as each contestant weighs his or her risks in the fight to live. I won't give away the ending, but I'll say this much: it's a great risk-balancing act.

Because of our risk thermostat, safety measures designed to reduce risk often produce unintended consequences, leading us to compensate by doing riskier things somewhere else. Risk compensation and the inherent tendency to move toward an acceptable level of risk is called *risk homeostasis*, coined by Dr. Gerald Wilde, professor emeritus at Queen's University, Kingston, Ontario, suggesting that we actively maintain a certain level of risk without consciously trying to do so. It comes naturally to us.

This one simple concept explains a great many things. We spend billions on levees and flood-control systems, but overall the number of flooding deaths doesn't go down. The reason? People feel the threat of flooding is reduced by the levees and move into flood-threatened regions. Floods may be less frequent, but the danger is greater when they do happen. We spend billions engineering our roads to be safer, but the number of traffic deaths per capita stays the same. The reason? We feel the roads are safer, so we take greater risks.

In his book *Target Risk*, Dr. Wilde explains how risk homeostasis works: "Risk Homeostasis Theory maintains that, in any activity, people accept a certain level of subjectively estimated risk to their health, safety, and other things they value, in exchange for the benefits they hope to receive from that activity (transportation, work, eating, drinking, drug use, recreation, romance, sports or whatever)."[17]

One way to look at risk homeostasis is to take the idea to an extreme. For example, to reduce the accident rate on a stretch of road to zero, you could just close it entirely. No cars, no accidents—no problem. But we don't drive on one stretch of road. Drivers will simply find a different route to get where they want to go, and they'll have accidents there instead. Wilde uses the analogy of a river that is blocked by dams covering two-thirds of its expanse. Does this mean that the flow of water will fall by two-thirds? No; the river keeps flowing because water will find a way to flow around, increasing the flow rate through the remaining portion.[18] And so it goes with drivers. If their risk is blocked in one area, they will simply move the risk somewhere else, keeping the risk approximately the same overall.

ARE YOU SMARTER THAN A CRASH-TEST DUMMY?

It might sound strange to suggest that all the engineering and regulation to make our roads safer doesn't reduce the number of deaths overall. Surely if a lack of seat belts is associated with X number of deaths per year, then getting everyone to wear seat belts will save X number of lives per year. It makes perfect sense, but unfortunately the data just don't seem to support this belief.

Large volumes of driving statistics are collected for the millions of cars and drivers on our roads, particularly regarding when things go wrong. Throwing all of these drivers and cars out on the roads is like doing a very large experiment where each driver is balancing myriad factors that influence his or her driving and, ultimately, his or her accident rate.

Statistics are often used to justify new safety measures, measures that usually look great on paper. Crash tests are another way of testing safety measures, putting dummies with sensors in cars and then smashing them into walls or into each other to see how well the measures work. When these measures are used out on real roads with real drivers, something strange happens. The safety measures don't save lives, at least not to the extent that's expected.

The rate of accidents and fatalities in SUVs is one example. People buy SUVs for a variety of reasons—although probably not for their fuel efficiency, I imagine. It's more likely that people buy SUVs because these big, heavy cars make them feel safe, secure, and relatively risk free. They (and maybe you) like the feeling of riding higher over the road than most other cars, believing these vehicles will protect their children as well. Indeed, in crash tests, the dummies in SUVs often fare better than the dummies in the lighter-weight vehicles they run into.

But crash tests fail to tell us what really happens out on real roads, in part because crash-test dummies aren't driving out here in the real world; humans are. If humans drove the same way all the time, regardless of the car they're in, then SUVs might indeed be safer. But real drivers vary their driving depending on the road and how they feel. If the road on which they're driving or the car they're driving makes them feel safe, then they'll relax and loosen up. And when we loosen up, we take more risks. Crash-test dummies might not know if they're driving an SUV or a Mini, but humans do. Humans know about the new safety measures being taken in cars, and we drive differently because of them.

To see how real drivers behave, Thomas Wenzel from the Lawrence Berkeley National Laboratory and Marc Ross from the University of Michigan Physics Department looked at the number of driver fatalities each year for the millions of cars registered in the United States.[19] Across the United States, every fatal crash is recorded in a database called Fatality Analysis Reporting System (FARS), which looks at 340 variables in each crash. The data are collected for a variety of reasons, but they just happen to be handy for looking at risk homeostasis as well.

After digging through this mountain of data, some of the results that emerge are not surprising. Drivers of sports cars, for example, pose a far greater risk to themselves than most other cars. No shocker there, really—the search for a thrill is part of the reason why people buy these cars. But what's surprising is that when Wenzel and Ross look at crash fatalities for SUVs, the vehicles were no safer for their drivers than other cars. There is some variation in risk for each class of vehicle, depending on the brand and model, but the safest of midpriced, midsized cars like the Camry are safer than any of the popular SUVs examined. And while SUV drivers had the same overall risk as drivers of smaller cars, they also created much greater risks to other drivers.[20]

The problem here is that drivers don't act the same once they get in an SUV. They are transformed into a different creature—an SUV driver—and they start driving differently as a result. Feeling secure in an SUV, drivers instinctively drive faster and with less care than other drivers not in SUVs. They don't intend to do this and probably don't even know they're doing it, but, nonetheless, the data says that they do it.

The risk-taking tendencies of SUV drivers have been measured in several ways. For example, researchers have gained insight into the minds of these drivers by seeing how they hold the steering wheel. When drivers are pressured and feel greater risk, they place two hands on the wheel at the ten and two o'clock positions, but, when relaxed, they tend to drive with one hand on the wheel, which is more comfortable but provides less control if any surprises pop up in the road. J. A. Thomas and D. Walton of Opus Central Laboratories in New Zealand found that SUV drivers they observed in New Zealand were more likely than other drivers to drive with one hand on the wheel, although when self-reporting their behavior, the drivers believed they were using two hands. The difference between self-reported data and concrete numbers from observers is not unusual; self-reported data is almost always wrong in all kinds of ways, reflecting more how we want to be seen rather than how we are really acting.[21]

Adherence to safety laws is another way to tell how secure or anxious drivers are feeling. When people don't feel at risk, they relax and break the pesky safety laws that, in most developed countries, require the use of seat belts. More recently, an increasing number of countries are passing laws to restrict the use of cell phones while driving because of accidents caused by distracted drivers, thus providing another law to break. Looking deeper into SUV drivers and their habits, researchers found that they wore seat belts less often and were talking on their cell phones more often than other drivers. In London, SUV drivers were observed talking on cell phones 8.2 percent of the time compared to 2 percent of the time for drivers of other cars.[22]

The impact of safety measures is less than might be hoped for because our minds automatically balance the increased safety by taking on greater risks somewhere else. If we feel we are protected, then we'll relax and take more chances, floating our risk level back up again. In addition to SUVs, another place we can see this curious behavior is in cars with antilock brakes (ABS).

Antilock brakes have been hailed as a great safety advance, keeping brakes from locking up, braking in a shorter space, and preventing cars from skidding out of control. All in all, these are good things; the brakes are an excellent feat of engineering. But the problem, once again, is that humans are at the wheel rather than computers or crash-test dummies.

One of the unique types of driver studied with regard to risky driving behavior is cabbies, including a famous study of cab drivers in Munich, Germany, examining the impact of antilock brakes on accident rates.[23] In the Munich study, the cabs were all the same except that one group had antilock brakes while the other did not. In theory, antilock brakes should reduce the number of accidents. But in reality, things didn't play out quite this way. Over a period of three years and over seven hundred accidents, the Munich cabs that had antilock brakes did not have fewer accidents than those without. They had more. The problem wasn't that the brakes were not functioning properly—the brakes were fine. The problem was the humans. The cab drivers with antilock brakes drove more recklessly because they felt they could, and they had an incentive to do so, picking up more fares and making more money by going faster. When accelerometers measured how cars were driven, cabbies with antilock brakes were driving faster, turning harder, and braking harder than other cabbies, taking more risks because they felt that the extra safety margin of the brakes allowed them to. By doing this, they canceled out any safety benefits they may have received from using antilock brakes. Observers placed in the cars, without the drivers knowing they were being observed, reached the same conclusion; drivers with antilock brakes were driving recklessly.[24]

It's not that we are doomed to take unnecessary risks on the road. I'm still wearing my seat belt, and I suggest you do the same. How we drive is up to each of us. Overall, however, the expectation that we can eliminate risk on our roads or anywhere else through engineering, education, or regulation alone doesn't seem to pay off.

So what will work? An alternative approach is to alter the benefits and rewards that motivate people to do risky things. Rather than changing the car or the road, we might have more luck if we change our mind, creating incentives for safer driving, for example. Such an incentive-based program to reduce driving risks was tested in California in the 1970s.[25] Rather than altering cars, roads, or regulations, the DMV program sought to alter the motivations that make drivers choose riskier driving, sending letters to 9,971 drivers who had caused collisions or committed violations in the previous year. The letters offered them a free twelve-month extension to their driver's license if they avoided any further driving problems, saving a few dollars a year. Another group of drivers did not receive the offer but had their driving behavior followed as well, to compare to the first group. Among drivers whose license was up for renewal within a year, the accident rate was 22 percent lower. And in the year after that, the drivers who earned the bonus had 33 percent fewer accidents.[26] The incentives were small, but the payoff was great. Why aren't we using this strategy more? Maybe it's hard to change how we think about risk and the strategies to deal with it, sticking with the prevailing notion that punishment is the best way to change behavior.

According to Dr. Wilde, "The accident rate and the incidence of unhealthy habits essentially depend on people's orientation towards their future. The more they expect from it, the more careful they will be with life and limb. If their expectations are low, they will try to find more immediate gratification of their desires, and do so at a greater risk of jeopardizing their lives." So while there's nothing wrong with seat belts and air bags, these steps alone aren't likely to protect us from our own inner drives and values.

BLOWING BUBBLES

Driving is a relatively easy place to look for risk compensation, but SUV drivers and cabbies are not the only ones taking risks in the world. Our jobs, our health, and our money are subject to the same influences, sometimes on a very large scale, as has been painfully clear in the financial crisis that rocked the world starting in 2007.

Handling money involves risk, and for financial professionals, handling risk well means making money. For some, a lot of money. Those working in the financial industry know that the risk is built into what they do. Risk cannot be eliminated, but it can be controlled and managed, often by hedging it, selling it, or otherwise moving the risk around in the market from someone who doesn't want it to someone else who does, for the right price. One way to view the markets is as enormous risk engines that weigh, value, and shift risks around the world.

One of the great advantages of hedging tools is that they allow businesses to reduce risks they don't want to take. Here is an example from James Angel, associate professor of finance at McDonough School of Business, Georgetown University:

> An oil company may have an old well that needs serious maintenance. This might be worthwhile at today's price of $100/barrel but not at $70/barrel. Because of the price risk, the company may not be willing to make the investment. However, by entering into a long-term forward contract to sell oil at a fixed price, the oil company has the assurance it needs to make the investment in extending the well's life. Thus, the producer can focus on doing what it does best, notably getting oil out of the ground.[27]

By managing the risk, the oil company can do the maintenance and pump the oil without worrying if the constantly changing value of oil will render the operation unprofitable.

On the other end of that risk transaction is another party, perhaps a speculator contracting to buy the oil in the future at a predetermined price. The speculator is taking the risk away from the oil company. If the price of oil is lower than the contract price, the speculator loses money. If the price of oil at that future date in the hedge is higher than expected, then he'll make money. "Note that the speculator is actually serving the socially useful function of risk bearing, just like an insurance company, and their [its] willingness to take on risk actually helps producers to produce more," said Angel.[28]

Knowingly buying a risk based on its perceived value is one thing. The problem in the financial crisis started when people thought that they could make risk disappear, like a rigged-shell game, but the risk was still there all the time, biding its time to come back and bite us all in our proverbial asset.

In the early 2000s, experts said we were in a new economy, one in which low unemployment was the new natural norm, powered by the productivity of the computer revolution finally hitting its stride. In October 2001, an article from the Federal Reserve Bank of St. Louis commented on the historically low unemployment rate at 4.5 percent: "Low unemployment rates, however, have become the norm following the longest period of sustained growth in the postwar period. Until recently, an unemployment rate below 6 percent was thought to be impossible without inducing rampant inflation."[29]

Meanwhile, real-estate values were rising steadily around the country and had been doing so for many years, posting annual double-digit gains. A rising number of people across the economic spectrum saw their lives and lifestyles buoyed by their rising asset values and readily available credit, buying homes and cashing out equity regularly with minimal documentation, even if their wages were flat or slowly sinking. Life was good. We were riding the bubble up and loving the ride.

While the bubble was on its way up, banks and investors learned a new way to make money—lots of it. Billions. In the old days, when a bank lent you money for a house, it would hold the note and you would pay it. The bank judged the risk that you would pay the money or not and decided if lending you the money was worth the risk. Then someone got clever. Rather than hold a mortgage and the risk that went with it, banks could sell the mortgage to someone else, getting the risk off of their books. Mortgage-backed securities went from a relatively small business valued in hundreds of millions of dollars to billions of dollars and then to trillions of dollars. If you hold a loan, then there's always a risk that some of the loans will go bad, but if you sell that loan to someone else as quickly as possible, usually within seventy-two hours, the risk practically goes away for the one originating the loan, almost magically, it seemed. And once the mortgage is sold, the originator can turn around and make more loans, cranking the handle on the money-making machine. Vast numbers of home loans could be packaged and sold to investment banks and investors, who would in turn insure the package against losses with American International Group (AIG). The more loans that are made and sold, the more money that is made. And many loans, like those from the Federal Housing Administration (FHA), were backed by the guarantee of the full faith and credit of the US government. In the case of the trillions of dollars' worth of loans held by the government-sponsored enterprises Fannie Mae and Freddie Mac, there was an implicit expectation that these loans wouldn't be allowed to fail. With virtually no risk to all the players involved, it seemed like nothing could go wrong while money was being made hand over fist on Wall Street, and ordinary folks got the homes and credit they wanted. Everybody was happy.

By cleverly packaging a variety of loans, the risk could be carefully managed, in theory. And it worked for several years, as long as home values were rising. But hiding in those packages, lower and lower quality loans were ticking away. When those creating mortgages no longer had the risk of holding them, they compensated for the reduced risk by making riskier and riskier loans and not worrying so much about all that pesky documentation or whether the loans could actually be paid. The risk was hidden, shuffled around, and forgotten, but it was still there, submerged and festering, growing and waiting to surface. And surface it did.

By 2006 there were signs of trouble, cracks developing in the shaky financial edifice. Home values stopped soaring, hovered in the air for a time, and then did the unthinkable—they started to fall. By 2007, home values seemed to be in a free fall, smashing through one barrier after another on their way down and triggering a chain of financial dominoes to start toppling. The shocks rippled throughout the economy and around the world, and the dominoes are still falling.

Subprime mortgage lenders could no longer sell their product as it started to smell bad, further accelerating the sickening collapse as credit dried up. From there, the contagion spread to homeowners who could no longer afford their payments, to employers who saw business dry up, to investment banks that handled the transactions, to AIG that insured the products that were quickly plummeting in value, then to the US government, and finally to all of us. By March 2008, investment bank Bear Sterns scrambled to stay afloat as rumors circulated that it was insolvent, until the banks turned their backs on it and the rumors came true. Fellow investment bank Lehman Brothers collapsed in September 2008 and drove the message home. Banks wouldn't even lend money to each other, each of them wondering if the other bank might blow up overnight, each bank wondering which of them was hiding the biggest and nastiest cesspool of toxic assets. The bubble had not just popped—it had blown up in our faces like a weapon of mass financial destruction.

Anthony Accetta, founder of the Accetta Group in Denver, Colorado, was there on Wall Street in the early 2000s as the bubble was inflating. Between 1999 and 2007, he was dealing with executives at the biggest investment banks in the world, including Lehman Brothers and Bear Sterns.[30] As a former Wall Street lawyer, a former federal prosecutor, and a private practitioner, Accetta has been working and interacting with financial institutions for many years from a variety of perspectives, including those at the heart of the mortgage crisis. In the 1970s, he had successfully prosecuted the biggest federal-mortgage-fraud prosecution in the United States. When he moved back to New York from Denver in 1999, he thought he was a savvy insider who had already seen how things worked. But Wall Street never fails to surprise. "I had been around and thought I knew a lot," said Accetta, "but I had no idea of the greed, the lack of integrity, and the willingness to walk on the edge. And if you go over, who cares."

His work at the Accetta Group involved doing risk analysis for firms. "We did background checks, but not the standard-operating background check. We did deep research into who companies were dealing with, the nature of the interaction, who the parties were, and the history of the companies we were dealing with. In 1998 to 2001 we were in the middle of the dot-com bubble. All of them [investment banks] had more deals than they knew what to do with. I helped them to cherry-pick and do the deals with the least risk, avoiding litigation and reputational risk from bad publicity."[31]

The investment banks make money from transactions like initial public offerings (IPOs), mergers, and acquisitions. "When the dot-come bubble burst though, for Lehman [Brothers], Bear Sterns, and all of them, the income from transactions fell off the table. Business fell apart. They needed the next big thing, what to invest in, and that's when they discovered mortgages," Accetta said in an interview.

The concept of securitization had already been there, but banks took it to a whole new level, hitting their peak in the mortgage frenzy that ensued. "The volumes in mortgage-backed securities from 1990 to 2007 increased from hundreds of millions, to billions, to trillions. I was working with presidents, CEOs, and boards involved in this stuff. And I observed a complete lack of risk aversion. Look at the nature of the real-estate business. Risk was not a factor in their thinking. It is the key to understand the housing collapse," remarked Accetta.

"From the very bottom to the very top, all of the risk was pushed upstream, and as long as there was not a catastrophic failure, the risks were minimal compared to the profits that could be obtained," He added. Along the way, in the years leading up to the crisis, there were voices that tried to express concern, but they were not popular voices. Accetta's voice was one of them, and speaking out on the topic today is probably not winning him many fans on Wall Street. For him, though, it's about doing the right thing, it's about going to sleep at night with a clear conscience.

"The whole process is fraught with risk, but the people who were controlling the money couldn't have cared less because they saw it as risk free," Accetta said. "I worked with the head of the fixed asset group at Bear Sterns, and in 2000, I told them, 'You guys can't do this. You need to have quality control in the mortgage-origination process.' They looked at me and they said, 'Are you crazy? That's a rounding error in our balance sheet.' The thought of losing a few hundred million didn't bother anybody."

The investment bankers didn't realize the risk they were exposed to wasn't just a few hundred million dollars but billions, more than enough to consume the whole company and spread like wildfire. "They thought the price of real estate would always cover any losses. But with the increased volumes in securities, they were taking it to levels nobody had ever seen. From 2005 to 2007 I was talking to clients, saying 'Let me set up a mortgage quality-control protocol for you guys.' The bottom line was that they could not have cared less."

Did anybody intentionally take on risks like these? Not exactly, it seems. But they were blind to the true size of the risk by the enormous sums of money they were making, by optimism, and by a failure to appreciate that while a total collapse seemed unlikely at any moment, it was almost guaranteed to happen eventually. That's the way author Nassim Nicolas Taleb's Black Swans work—they're predictably unpredictable. "I don't think they believed that the whole housing industry would collapse," said Accetta. "They thought that AIG would be there to cover things if there were any problems. Even I never imagined that it could be as bad as it is. The phrase I use is *conscious avoidance and reckless disregard*. They were recklessly disregarding the facts."[32]

Before the bubble popped, when things were looking nice and bubbly, I guess one could be excused of having a bit of irrational exuberance. Even Alan Greenspan drank the bubbly Kool-Aid. But having lived through the last few years, we should know better now, shouldn't we?

Has anything really changed though? Uh-oh, don't tell me . . .

"No," said Accetta flatly. "We've had some legislation, and frankly it's all bullshit. Right now, everybody is waiting for the whole thing to shake loose, and sooner or later it will, and then the whole thing will happen all over again. We've seen bubble after bubble after bubble, and what has happened? Look at the dollar value of the bubbles that have collapsed one after another and ask yourself, 'Why is this bigger than the one before?' Nobody has taken the step to say 'Wait a minute, boys, this won't work.'"

With nothing standing in the way, the fear will subside, credit will loosen up, and money will start chasing assets again, with those in the middle skimming a chunk as the next bubble starts to inflate. "Without a deterrent, you incentivize, and they'll do the same thing again. There has not been one single prosecution of a major Wall Street figure."[33]

So what's the next bubble going to be? Chinese real estate? Credit-default swaps? Tulips?

The overall derivatives market totals hundreds of trillions of dollars, $700 trillion or so, far greater than the combined assets of the whole globe, which ring in at about $150 trillion. If you own Greek government bonds, you can buy derivatives that are designed to reduce the risk you face that the bonds will drop in value. In addition, you can get insurance to cover the derivatives, making the whole transaction feel seemingly risk-free. The risk goes in, but it doesn't come out. Or so it seems.

In practice, derivatives are so complex and hard to track that nobody fully understands how the derivatives really work. The risk is hidden and shuffled around so effectively that even the professionals lose track of it. But the risk is still there and even magnified because when risk is hidden and people feel safer, they find new ways to generate risk, going out on limb farther than they would have before. Rather than reducing risk, the net effect may be to increase it. While there were some calls to regulate the derivatives market and defuse this huge pool of lurking risk right after the onset of the financial crisis, nothing has really changed. The business is just too profitable to touch.

Financial players don't make money by playing it safe, at least not in the short term. Their incentives push them closer and closer to the edge, with derivatives, mortgage-backed securities, and credit-default swaps helping them to scooch out on the ledge even farther, teetering over the side. Warren Buffet has called derivatives a "time bomb" and a "mega-catastrophic risk."[34] With the risk hidden and insured, players in the derivatives market feel secure, emboldening them to place even riskier bets. The banks were like the speeding taxi drivers with anti-lock brakes, emboldened by protective measures to move forward more carelessly. Sooner or later, they were bound to crash.

The current financial crisis is not the first, and it won't be the last. The up-and-down cycles are not just fluctuations in asset values but also fluctuations in how people see risk, oscillating slowly around a set point. When times are good, little risk is seen and people go out on the limb farther and farther. When times are bad, the risk-perception pendulum swings the other way and everything starts to look risky, keeping the value of assets low because of the perceived risk. The assets involved will differ, and the regulations may change, but the psychology doesn't.

"We are social animals who are biologically programmed to stick with the herd," said James Angel. "If you deviate from the tribe, the sabre-tooth tiger might eat you. If other people are doing it, it must be the right thing because maybe they know something we don't. This can lead to a society-wide groupthink that ignores certain risks. This happened big time in the subprime meltdown. With so many presumably smart lenders making subprime loans, it looked like they had to have figured out a way to manage the risk. So maybe I should imitate them and make money doing it."[35]

BUT WAIT, THERE'S MORE

Taking vitamins is another example of risk compensation. We're told that many people do not receive sufficient vitamins and minerals in their diets due to a lack of proper diet, among other reasons. Vitamins do indeed play key roles in the body, and people often have poor diets. And we love the idea that taking a pill or a supplement can quickly fix all that.

There might not be many cases of scurvy (resulting from a vitamin C deficiency) or rickets (caused by a vitamin D deficiency) encountered by doctors today, and goiters caused by a lack of iodine are not commonly seen in the Western world. However, some people do suffer from insufficient vitamin intake. Consider folic acid, a B vitamin. In some pregnant women, lower-than-normal folate levels have been related to birth defects, a situation that is easily resolved by taking prenatal vitamins, which contain folic acid. Along the same lines, adults who get little sun and don't eat dairy products or fatty fish are still at risk of getting too little vitamin D.

One-half of Americans take vitamins and supplements of some sort or another, helping these products to grow into a $27 billion industry.[36] And yet, studies to test the health benefits of vitamins and supplements often find little or no impact. While supplement sales have soared in the United States, our overall health has not. One reason might be that most people eating a reasonably healthy diet probably get enough of the nutrients and vitamins they need and don't need the supplements. Another problem might be that, in the real world, we don't keep our diet constant when taking supplements. We risk compensate.

Researchers Wen-Bin Chiou and his colleagues showed this in a study in Taiwan.[37] Volunteers were all given a placebo pill (a pill with no active ingredients), but some of the volunteers were told it was a multivitamin tablet that was being studied while others were simply told it was a placebo. Afterward, the volunteers in the study were given surveys to indicate how much they enjoyed either unhealthy activities that produced an easy immediate reward, like sunbathing or going to wild parties, or healthier activities that were not as easy, like running or kayaking. In addition, participants were given a survey to measure their feelings of invulnerability and were provided a coupon to have either a free, healthy, organic lunch or a lunch at the all-you-can-eat buffet, loaded with the less healthy fare toward which we often gravitate.

Sure enough, those who thought they had taken the vitamins had a shift in their preferences, going more for the wild parties and less for running. And rather than going for a healthy, organic lunch, they fell into the gravitational pull of the buffet. Having taken what they thought was a multivitamin, they felt they had filled up on invulnerability in the health bank and could spend some of those health points by slacking off somewhere else.

The volunteers in the study are not the only ones who do this because it's human nature. If we exercise, then we feel like we can reward ourselves with a treat or with junk food. I did it myself today—I went for a six-mile run and afterward went for lunch at McDonalds. If we take the multivitamin, we feel secure and figure that we can ease up on the rest of our diet and eat whatever we want. "Hence, people who rely on dietary supplements for health protection may pay a hidden price: the curse of licensed self-indulgence," Chiou and his colleagues concluded. A curse indeed.

A similar story may be behind our consumption of diet sodas. Consumption of diet sodas has soared in the United States, with Americans each guzzling gallons of the stuff, on average. The idea is to consume less sugar by drinking diet soda instead of the sugar-loaded kind, and yet even with the slew of diet drinks, the obesity epidemic continues unabated. If the calories in sodas are really the problem, loading people up on gallons of high-fructose corn syrup and thousands of empty calories, then surely diet sodas would help, wouldn't they? And yet, they don't. In fact, they seem to make things worse. Data from epidemiologists at the School of Medicine at the University of Texas Health Science Center–San Antonio found that drinkers of diet soda actually gained more weight. Those who described themselves as frequent diet-soda drinkers, drinking two or more servings per day, had a 500 percent greater increase in their waist circumference than those who did not drink diet soda.[38]

There may be a biochemical component to this, with the artificial sweeteners doing something in our body that nobody imagined, but it probably involves risk compensation as well. Your mind does the quick calculation and decides that if you are saving some calories over *here* by drinking diet soda, then you can have a little more of something else over *there*. And since we are incredibly bad at tracking and judging our own eating habits, we end up overcompensating and eating even more than we would have without the diet sodas.

Skiing provides another example of risk compensation. Skiers make mental tradeoffs on the fly regarding their safety as they're headed down the slopes. Downhill skiers hurtling down a slope can reach speeds of up to fifty miles per hour (and even higher), and they are not alone on the slopes. On your way down, it's hard to make a sudden stop; even skilled skiers can run into trees, rocks, and other skiers or snowboarders, causing them to lose control. This can have serious consequences, death being one, with hitting your head as the most likely outcome. Some have suggested mandating that skiers wear helmets while skiing. A helmet may not save everyone, but it probably would not hurt and would protect many. Besides the problem of not looking very cool, it is again likely that, at the end of the day, helmets do not do as much as we hope because we outwit the helmet.

Dr. Mike Langran is the president of the International Society for Skiing Safety. While an advocate of wearing helmets, he also knows that helmets can do only so much. There's data that even when helmets are worn, they don't always work. "There is some research that indicates helmet wearers take more risks," he said. Wearing a helmet, we feel safer so we go faster, take more risks, and get into more trouble, negating the benefits of wearing a helmet.[39]

Protection against sexually transmitted diseases is yet another fertile ground for risk compensation to exert its influence. A variety of means of reducing the risk of HIV transmission are available, from condoms to antiretroviral drugs. The more that people are aware of these and feel less at risk, the more likely it is that they will relax and engage in unsafe sex, increasing their risk again.[40] Similarly, some worry that girls who receive the HPV vaccine may overestimate the protection it provides. The vaccine protects against the human papilloma virus, associated with an increased risk of cervical cancer in women. The virus is sexually transmitted, and, like most vaccines, the protection is not absolute. Safe sex is still a good idea. But the connection is hazy for some girls. Of 339 girls surveyed, about one-quarter believed that, with the vaccine, they would be protected against sexually transmitted diseases of all sorts.[41] Not just HPV, but STDs in general. With this belief in hand, the girls may be more likely to not practice safe sex with condom use and, in the process, to expose themselves not just to HPV but potentially also to syphilis, gonorrhea, and HIV. While some have raised concerns that girls may practice unsafe sex if they have beliefs like this, Dr. H. Hunter Handsfield, professor of medicine at the University of Washington

Center for AIDS and STD in Seattle, commented that "if anything, it's a fairly pleasing result that it's only 24 percent."[42] Probably not that pleasing to the parents of the 24 percent, though.

OCCUPATIONAL HAZARDS

Some risks come and go, while others have a constant presence, like risks on the job. Some jobs require constant vigilance to prevent accidents or other problems from happening. Air-traffic controllers, airline pilots, nuclear-plant operators, and security guards are just a few examples. These workers are required to pay perfect attention for long periods of time, which is quite hard to do. And efforts to make it easier might actually be making it worse.

While many passengers find air travel frightening, flying is all too routine for flight crews as automation does more and more of the work. Although the rate of airline crashes has steadily decreased, 50–70 percent of accidents are attributed to human errors, mostly on the part of the flight crew. While automation is making flights safer, it may also be contributing to inattention and lack of preparedness by flight crews who get used to having automation do the work. "The more automated things get, the more difficult it gets to spend 16 hours at a time in the cockpit and stay engaged," said Paul Rice, an airline captain and a vice president of the Air Line Pilots Association.[43]

Some occupations require long periods of vigilance against significant risks, often at odd hours of the day. Air-traffic controllers, security guards, and nuclear-plant operators have boring jobs with a great deal on the line. Overall, their job is to keep things as uneventful and as boring as possible. But the boredom itself can become a risk. They know they have important work to do with a lot at stake, but over time it still becomes a struggle to do the work, or even to stay awake, no matter how much coffee is consumed. On multiple occasions, air-traffic controllers have fallen asleep in the tower while on night shifts, leaving planes' pilots to sort out landings on their own.

Being a pilot is one of the top-ten riskiest jobs, and air-ambulance helicopter pilots specifically have one of the highest risks of all, higher than crab fishing, logging, and ice-road trucking.[44] In 2008, there were thirteen crashes of medical helicopters involving twenty-nine deaths.[45] A report commissioned by Bell Helicopters described several potential factors for safety problems, including pilot fatigue, lack of instrumentation, lack of training, and reimbursement policies. Another likely factor is that inexperienced pilots flying in difficult conditions (and doing it routinely) make riskier choices than they should.

Steve Greene has been flying air-ambulance helicopters for over twenty-eight years.[46] He got started flying in the military when he was eighteen years old, flying helicopters in Vietnam, and got involved in the air-ambulance industry as it first started to grow in the United States. Greene works in the Ukiah area in Northern California, about a hundred miles north of San Francisco. The work involves transportation of patients for medical treatment, often in response to a 911 call, landing on a front yard, in a field, or at an intersection, if necessary. He's worked with patients suffering from heart attacks, stabbings, electrocution, injuries from car crashes, or just about anything else humans can do to badly hurt themselves, except perhaps a broken heart (see ch. 3 for this one).

Often air-ambulance pilots spend a great deal of time waiting for the phone to ring at any second, like firemen or policemen. When they're on duty, they stay in living quarters, with a bedroom and a television, waiting and waiting. Then when it's time to go, there's hardly a chance to worry, although there's plenty to worry about if someone wanted to. Accidents are all too common.

"I always tell people, I know more guys who killed themselves in crashes than retired from the business," said Greene. What happened to pilots who have ended their lives in crashes? "It's all kinds of different things," said Greene. "Often it's controlled flight into terrain. Pilots fly into the ground because of bad weather or darkness or both, trying to maintain visual conditions when it's not possible to do this."

He knows the statistics for air-ambulance pilots, but he doesn't feel at risk himself. "From my perspective, it's not that risky. I'm not all that brave," he said. Greene has many years of experience, is aware of the challenges, and is out to do a great job in challenging conditions. "I feel that I'm well prepared to do what I do, given the tools that I have available, [and] that things should work out well for me. Even though it's a risky occupation, the odds are still in my favor."

Even the riskiest of occupations, like air-ambulance pilot, can become routine. "I flew in combat, flying every day, and soon that becomes the norm," said Greene. "If you're not vigilant, you let your guard drop, and get in trouble. . . . At any moment of the day or night while we are on duty, we are called upon to fly somewhere, with little preparation—speed is of the essence in this field—then land, assess our patient, and fly [him or her] to the appropriate medical facility. That is our routine. We attempt to calculate the risk before and during each flight, but hundreds of us and sometimes our patients have died when we failed at assessing risks or by making mistakes."

There is at least one safety benefit of air-ambulance flying that Greene noted when I talked with him: at least when he's picking up patients these days, generally nobody's shooting at him, unlike his days in the military.

Ignoring common risks may be dangerous at times, but it can also be useful and may even be necessary. In the world of psychology, desensitization therapy uses repeated safe exposure to spiders or airplanes to decrease the sense of risk and eliminate fears. If you have to drive every day, it's best to do it without being panicked behind the wheel. "Our brains are wired to work on automatic pilot in everyday life," said Prof. Gerard Hodgkinson, director of the Centre for Organisational Strategy, Learning and Change at the University of Leeds in the United Kingdom. "If we considered and analyzed the risks involved in every permutation of every situation, we'd never get anything done."[47]

After the trouble passes, the level of alarm resets back to normal most of the time. When a lion attacks at the watering hole, there is, for a time, pandemonium among baboons, but the pandemonium is temporary. Eventually the predator gives up or gets its meal, and life goes back to normal for everyone that is left.

As I was writing this chapter and looking over the statistics for auto accidents, I found myself on the freeway one day, taking the ramp from 52 West to 5 North in San Diego. As I drove, I started looking around at the other cars near me on the road, glancing at the drivers more attentively than I normally would. It was rush hour, around eight o'clock in the morning, but the traffic was moving quickly with cars packed close around me like a menacing school of barracudas. A few feet away in every car, someone was talking on the phone, brushing her hair, and fiddling with who knows what, doing everything but paying attention to the road. I am sure they were doing the same thing every other day when I was less aware of it, but this time I was thinking, "Forty thousand deaths a year," painfully aware of the asphalt sailing past a few inches beneath my body, cars closing in around me, tight and fast. I broke out in a cold sweat, thinking "How the heck do we get away with this every day?" One wrong move, one blown tire, one sneeze, and—*kablooey*. Chaos could erupt at any moment, a hair's breadth away.

I moved on, and the feeling quickly faded. I got to my destination and went about my business. "Nothing bad happens to me." There's a reason why we live with this basic assumption—because it's useful. We need to let go of our fears to get on with the daily business of living. Often fear is a bigger threat to us than the many risks we face, leading us toward a variety of ways of fighting fear. One is denial, as we've seen. Another is by gaining control of our lives, or at least feeling like we're in control, as we'll see next.

CHAPTER 6

LOSING CONTROL AND GAINING FEAR

Small Aircraft, Fear of Riding Along, and Conspiracy Theories

THE QUEST FOR CONTROL

Among our many needs as humans, some are quite obvious. We need to eat, breathe, and drink; anything that blocks our ability to get these needs met sends our internal riskometer skyrocketing. Like it or not, we also need to sleep. Many would include a need for social contact as an equally fundamental part of being fully human (see ch. 8), and our need for sex and love seems hard to deny (see ch. 3). In addition to these basic needs are others that might seem less obvious at first, and one of these is a need for control. Control is not just for control freaks. We all need it one way or another, and we all feel a lack of control as a very risky proposition.

Living things of all stripes have a strong instinct to survive, and taking control helps them do this. Feeling like we are in control of our lives and our surroundings makes us feel safer and makes the world feel less threatening, while losing control makes us feel anxious and depressed. "One of the most prevalent fears people have is that of losing control," said psychologist Elliot D. Cohen. "This is the fear that if you don't manage to control the outcome of future events, something terrible will happen."[1]

While we feel compelled to seek control, our lives are swept by events, trends, and influences completely out of our control. Don't look now, but if you pick up a newspaper or look out the window, it's not hard to find massive events sweeping the globe, the street, or the Internet, and they're carrying us along with them (I told you not to look). The weather, the stock market, and the actions of billions of people are all pretty much beyond your control. Our jobs often seem out of our control as well, swept up in the machinations of the global economy and corporate boardrooms. Even those close to us, like our spouses, children, and friends, can seem at times like wild animals, bent on their own courses, no matter what we say or do. It's like they've got minds of their own or something.

"The less we feel in control, the less willing we are to take a risk," Paul Slovic, professor of psychology at the University of Oregon and president of Decision Research Group, said when I talked with him.[2] "Imagine a thought experiment where you're slicing a loaf of bread," he tells me. I see myself in the kitchen, knife in hand, with some bread on a cutting board in front of me. "Where would you put your hand on the bread relative to the knife blade?" I often hold things close to the knife blade when cutting food in the kitchen, probably closer than I should. I don't cut myself, though. (Well, not often anyway.) "Where you would you hold the bread if someone else was holding the knife?" That's an uncomfortable image, probably because no matter who the other person is or how much you trust them, they have their own brain and nervous system. You never know what the other guy is thinking or how much wine he had. I'd definitely inch my hand farther back on the bread, away from the knife blade, if someone else was holding the knife. "The more control we feel, the more willing we are to take risks."[3]

This quest for control is not new. Ancient cave drawings of

animals and hunters may represent an early effort by ancient man to gain control of his world. If he painted the animals, perhaps he felt he might gain some control over them and they would then be more likely to cooperate with his hunting. Or maybe it was just the prehistoric version of PowerPoint. It's hard to say for sure what these people were thinking based on cave drawings alone, since their thoughts left no fossils.

One of the benefits of developing technology is that we have at least *some* control over our immediate surroundings. We can usually have light, water, and food almost anytime we want them, at the flick of a switch or the twist of the faucet. Technology has its limits, unfortunately, but if we still feel powerless at times, imagine how early humans must have felt when armed with little more than simple stone tools and figures painted on a wall.

LEARNING HELPLESSNESS

One sign that the quest for control runs deep in our blood is that other animals, like monkeys or mice, have the same hunger for control. If you put a monkey in a chamber and allow it to push a button and subsequently receive a treat, it will quickly learn which button to push. In addition to getting something to eat, it learns that through its actions it can gain control and feel better about the world, reducing its stress level compared to other animals that lack control. You can measure a monkey's stress level by how it behaves, its overall health, and the level of stress hormones, like cortisol, present in its blood. The more in control monkeys feel, the lower their stress, and the healthier they are overall.

The opposite of control is helplessness, a state of great anxiety and stress because of the risks faced when one is helpless. Psychologist

Martin Seligman studied learned helplessness in dogs, a condition where animals that normally show a healthy ability to avoid problems are trained to believe that they have no control to help themselves.[4] Seligman took three groups of dogs and trained them three different ways, putting a harness on each of them and putting them in a chamber, one at a time, where some of them received mild electric shocks. (The shocks were unpleasant but did not injure the dogs.) In the first group of animals, each wore a harness and was put in the chamber—but that was it. No shocks; no problem. In the second group of dogs, each wore the harness and received mild electric shocks in the chamber, but they could stop the shocks by pressing a button with their nose. They had control. In the third group of dogs, each wore the harness and received the shocks, but when they pressed the button, it did not alter the electric shocks. They had no control.

Later, the animals were taken to a different lab and placed in an enclosed area with two chambers separated by a low wall. On one side, they would receive shocks again, but they could jump over the wall onto the other side to avoid getting shocked. The jump from one side to the other was a small one, and the dogs from the first two groups quickly figured out that they could get away from the shocks by jumping into the other chamber. The third group of dogs, which had received random shocks they could not stop, did not figure out how to jump to safety in the new chamber. Even as they received shocks, they did nothing to get away. They had learned previously that they were helpless to prevent the shocks, so even though they could have easily jumped away to avoid the shocks, they stuck with the helplessness they had learned and did nothing. They had learned how to be helpless, the opposite of control. Believing that there was nothing they could do, the poor dogs lay down and gave up, resigned to their fate in an apparently cruel and chaotic world.

People often go through the same thing, one way or another. When people experience bad things, particularly if they experience them repeatedly when they are young, they sometimes feel that there's nothing they can do to help themselves, that their fate is beyond their control. Being abused physically or emotionally sends some victims into tailspins of helplessness that can flavor their views of the world for the rest of their lives. Fortunately the opposite can be learned as well: learned optimism, which is the focus of Dr. Seligman's more recent work.

THE POWER OF CHOICE

The way we see risk rarely parallels the statistical risks we face, and having or lacking control of a risk is one way to amplify how risky something looks. When we have control of something, it feels far less risky to us than when we don't.

While most of us think skydiving is a risky thing to do, others love it. It might not be for everyone, but unless they are suicidal, those who do it have decided that it's an acceptable risk. This isn't to say there aren't accidents; in 2007 there were eighteen skydiving deaths out of 2.5 million jumps in the United States.[5] This might sound like very few accidents and could be used as evidence in support of skydiving, if you love the sport. Alternatively, it might sound like a good reason to stay away from skydiving if you hate the idea of jumping out of an airplane. What makes the risk acceptable to those who decide to do it is that they like doing it and are in control. Nobody is getting pushed out of an airplane (hopefully)—they are making the choice to jump out and, because they have control over the jump, it seems okay to them.

Chauncey Starr was a pioneer in the field of risk perception, looking at some of the key influences on how risky we think things are. He proposed that two of the biggest factors that shrink or swell our sense of risk are whether we like something and how much control over it we have. If we like something, we find a way think it is safe; if we dread something, it feels risky and we choose to stay away. Starr estimated that if you don't have control over something, it feels a thousandfold riskier than it would otherwise.[6]

Imagine, for example, that you love eating fugu, a Japanese dish made from the puffer fish. Fugu contains the deadly neuro-toxin called *tetrodotoxin*, one of the most potent and deadliest toxins known. A dish prepared from the puffer fish is considered a rare delicacy in Japan, served only by specially certified chefs carefully trained to avoid poisoning clients. If you eat fugu that is not cleaned and prepared properly, the tetrodotoxin it contains induces paralysis and you stop breathing. Too much of that, and you'll never eat fugu again. It has happened to fugu lovers more than once. Fugu lovers enjoy getting just enough toxin to get a mild tingling and numbness around the lips and mouth (while still breathing, of course), some-thing that only a fugu aficionado could love. To many of us, the pos-sibility of death makes a fugu meal feel too risky, not really worth it. Personally, it's hard for me to imagine how good this fish could be to risk paralysis and death; I'll stick to tilapia and salmon. But if you love fugu enough, you'll downplay the risk in your mind and find it to be acceptable.

As a leading figure in research regarding risk perception, Paul Slovic has expanded on these observations. In looking at our percep-tions of everything from nuclear accidents to artificial sweeteners, Slovic sees control and dread as two of the most important factors that shape the risks we perceive. The less control we have and the more we dread something, the more risk we feel it presents.

People find radiation from a potential nuclear accident particularly risky because the threat is invisible, out of their control, and it can cause cancer, which scores high in dread points. But not just any amount of radiation seems to be associated with an increased cancer risk. Studying the impact of radiation in Japan after World War II, a dose of 50 millisieverts (a unit of radiation exposure) in a year was found to be the threshold beyond which there was an increased risk of cancer. This is by no means accepted by everyone, but this is the figure typically used by regulators. Because there is always some amount of radiation in the world, emitted both from the sun and from the earth beneath our feet, Americans get about 6 millisieverts' worth of exposure each year, no matter what we do. Medical scans are now the biggest source of radiation exposure for most Americans, in the form of MRIs, CAT scans, and x-rays; a patient can receive 10–25 millisieverts from a single study, an appreciable fraction of the dose regulatory officials worry about. Most of us think little about moderate doses of radiation at the dentist's or doctor's office, but we worry more about the risk posed by a fraction of the dose from a source that's out of our control, like a nuclear accident. The difference is that patients choose to be scanned and believe the scan to be helpful, but a nuclear accident has neither of these attributes. Nobody chooses to be around when a nuclear power plant spirals out of control.[7]

When the nuclear reactor in Fukushima, Japan, was overwhelmed by the earthquake and tsunami in March 2011, it triggered one of the worst nuclear emergencies in the world to date. The plant had been built with a seawall twenty feet high for protection, but the tsunami reached the coast as a wall of water forty-five feet tall, topping the seawall and disabling the power supply and backup generators used for cooling the reactor. The core in multiple reactors at the site melted, and radiation spewed into the air and water. Even as the Japanese struggled to deal with the havoc wreaked by the earthquake and tsunami, the man-made reactor disaster was added to the trouble, leading to the evacuation of many in the region and a long struggle to bring the reactors under control.

The impact in Japan was great, but the Japanese were not the only ones affected. As radiation from Fukushima drifted across the Pacific Ocean, people on the West Coast of the United States worried about radioactive fallout when trace quantities of radiation were detected in the atmosphere and in water samples taken from California in March 2011. California is a long way from Japan, and the radiation levels were so low by the time it crossed the Pacific that, while detectable, the radiation was at far lower levels than would pose a hazard. However, lacking control, the distance and dilution did not feel like enough protection to many. The fact that the radiation was there at *any* level, no matter how low, was enough to sound the risk alarm because people have little control over the air they breathe. Potassium iodide pills sold out as worried people took steps to protect themselves from the perceived risk of thyroid cancer, even if taking the pills comes with its own risk.[8] But people buying iodine pills were doing more than taking pills—they were taking back control.

Fukushima was only the second grade-seven nuclear accident ever, which is the highest ranking in the International Nuclear and Radiological Event Scale created by the International Atomic Energy Agency (IAEA), but it was not the largest. The first was at Chernobyl in April 1986, which released ten times more radiation than Fukushima. The problem at Chernobyl started during a technical exercise that was intended to eliminate a safety problem with cooling systems if power was ever interrupted. In efforts to run a test to solve this problem, numerous safety systems were intentionally taken off-line, and the plant was placed far out of safe operating parameters. The test was running behind schedule, delayed from the day shift to the poorly prepared night shift. When alarms started to sound during the test, they were ignored. A first explosion caused by steam ripped open the reactor vessel, and a second explosion with the force of ten tons of TNT ripped apart the core, exposing graphite, which started to burn and spew radiation, eventually releasing almost six tons of nuclear fuel into the atmosphere.

In some areas of the Chernobyl reactor, workers received a lethal dose of radiation within minutes. Ignoring chunks of graphite on the ground and a burning reactor, and without a working dosimeter that could function at such high radiation levels, they assumed things were not too bad. Now that's optimism for you. ("Nothing bad happens to me!") When the firefighters showed up, they had no idea what they were dealing with and had no protective gear for dealing with radiation. Some even picked up the chunks of debris, wondering what it was. Many firefighters later died from their radiation exposure.

Outside the plant, nobody was informed of the accident at first. The day after, the nearby city of Pripyat, where many of the plant's workers lived, was quietly evacuated, with residents being told they could return in a few days. The city remains evacuated today. Russia only spoke publicly about Chernobyl days later, after radiation alarms in Sweden detected the released radiation.

After Chernobyl, large swaths of land in the region were eventually evacuated, and much of the region is still radioactive and off-limits. Pripyat is a ghost town, with trees and bushes growing up through the roads and buildings, reclaiming the land. The schoolbooks and desks are strewn, just as they were left in haste. An abandoned amusement park lies rusting, bumper cars resting where they were left that day. It's like a glimpse of a postapocalyptic world.

The risk of low-level radiation can be hard to know for sure, even after witnessing the unfortunate experience of Hiroshima's exposure from the nuclear weapons detonated in World War II. It's been over twenty-five years that people have been exposed to low-level exposure from Chernobyl, and there has indeed been an increase in some cancers in the region. Of those who worked on-site or fought the initial fires, about fifty workers exposed to high-level radiation at the time eventually died. In the broader population, estimates of cancer deaths vary hugely, from four thousand cancer deaths estimated by the United Nations to totals reaching one hundred thousand deaths or higher estimated by environmental groups.[9] In children and teens in the region, about four thousand to five thousand children contracted thyroid cancer, and most survived, since this cancer responds well to treatment. Nuclear accidents on this scale are greatly feared by many for a reason.

Nuclear accidents have many of the traits that put us on edge. We have no control; the risk is unseen, potentially catastrophic, and is dreaded, associated with very unpleasant consequences like cancer. The risk of cancer is not the only problem facing those in these affected regions. For residents living in the area of Chernobyl, the sense of dread, of resignation to their fate, has proven a crippling problem because they feel the risk is out of their control. They are so pessimistic about their prospects for the future that, for many, the pessimism has become more of a problem than the actual radiation. When representatives from the IAEA went to the Chernobyl area three years after the accident to look for health problems, they saw little in the way of problems related to radiation exposure, but they did find pervasive anxiety. Three years might not be long enough to say much about the long-term impact of radiation, but even in areas where significant radiation exposure did not occur, the emotional impact of anxiety could be seen readily (and was greatest in mothers of small children, according to one study[10]). Residents felt helpless, their lives removed from their control by the accident.[11]

PLANES, STRAINS, AND AUTOMOBILES

One of the places where we acutely feel a need for control is on airplanes; people fear commercial jets, even though they are far safer than the freeway, because passengers lose control of their fate once they step on board the plane. Pilots, on the other hand, have a strong sense of control, a useful trait when your life and the lives of others on board with you depend on how well you can keep control of an airplane cruising miles high in the air.

On March 27, 2012, after the copilot locked him out of the cockpit while in the air on a JetBlue flight, the pilot ran up and down the aisles, shouting "They're going to take us down!" and talking about a potential bomb on board.[12] No doubt this was disconcerting for passengers, belted into their seats at thirty thousand feet in the air, with their inherent need for control screaming at them internally. After passengers subdued the errant pilot, the plane landed without further incident, his problems attributed to a medical condition.[13]

When it comes to how people view control, there are two types of people: those who feel like they are in control, and those who think events are controlled by the rest of the world. Psychologists call the difference between these the *locus of control*, grouping people into those with an internal locus of control (feeling that they control things), and those with an external locus of control (feeling that control comes from everywhere else). When pilots are tested, they typically have a strong internal locus of control, which is also correlated with having fewer accidents among military pilots.[14]

While people are afraid of commercial airlines because they don't have control, and accidents tend to be spectacular and widely televised; most aviation deaths (94 percent) are from smaller aircraft.[15] While people often note that you're safer flying rather than driving, this is only true for commercial airlines. With general aviation (small, private airplanes), the rate of death is much higher than with commercial airlines. So high, in fact, that general aviation has about eighty times more deaths than commercial aviation per time in the air.[16]

Flying in a small, private airplane is nerve-racking for many of us, but pilots of small aircraft aren't all that worried. In fact, they feel less at risk than the numbers warrant, perhaps because pilots are prone to overconfidence and because they are in control. When general-aviation pilots were surveyed and asked to rank the risk of various flying conditions, the pilots who displayed the lowest sense of risk and the greatest self-confidence were also those who made riskier decisions about how to handle the scenarios described. David Hunter and colleagues in New Zealand looked how the relationship between how pilots perceive bad weather and their risk-taking behavior in the air.[17] Pilots who had a lot of experience with bad weather started to view it as less risky and started to exhibit riskier behavior.[18]

Cars have their share of risk too, of course, but the view of risk in a car can look surprisingly different from the driver's side and from the passenger's side because the driver is the one in control. If you've ever been in the car with someone who was uncomfortable unless they were behind the wheel, squirming with physical discomfort as a passenger, you know what this is like. And no, I really wasn't going all that quickly, or at least it didn't feel that fast to me while I was driving. While it seems like passengers in a car have an easy time of it, the driver in a car pretty much calls the shots, controlling the risk of an accident as well as controlling where you go, while the passenger is pretty much at the driver's mercy.

In 1993, Frank McKenna at the University of Reading asked participants in a study how much of a risk of getting in an accident they felt if they were either a passenger or a driver in a car. McKenna questioned them about a range of scenarios in which the driver would have either a high degree of control or little control.[19] The questions about a low-control scenario included, "Compared to the average driver, how likely do you feel you are to be involved in an accident which is caused by oil on the road?" There's not much the driver can do about the oil on the road or its impact on the car, so a question like this is a way of testing how much risk they feel when they lack control. Scenarios where the driver would have more control included things like "Compared to the average driver, how likely do you think you are to be involved in an accident in which the vehicle you are in is changing lanes?" Those asked about all of these scenarios reported feeling much greater risk as passengers, when they lacked control, and also greater risk in situations when they lacked control as drivers.[20]

Our need for control seems likely to be related to the phenomenon of backseat drivers, those who are passengers in a car and yet provide a running critique of the driver's skills. While many passengers might feel anxiety and at risk, as seen in McKenna's study, backseat drivers are so anxious that they attempt to gain control of the situation verbally. (Note: reaching out and grabbing the steering wheel is probably not a good idea.) Through their words and increased sense of control, they reduce their sense of risk.

Having a feeling of control might be a good thing even if this feeling is an illusion. McKenna writes, "Exaggerated beliefs in control and unrealistic optimism can be associated with higher motivation and greater persistence in situations of objectively poor probabilities of success. The overall result is more effective performance and in the end greater success."[21]

We generally think that the well-adjusted among us are those who see the world most realistically, but this might not always be true. The well-adjusted are often the most optimistic, but that optimism isn't always borne out by reality. The pessimistic and depressed individual may have a more sober and realistic appraisal of the risks and opportunities facing them, but this does not necessarily help them to cope with these challenges. Being optimistic, even if the situation doesn't warrant it, and not seeing risks allows the optimists to take action and continue forward at times when the pessimist doesn't. Feeling in control and less likely to give up, optimists may overcome the odds and actually succeed. The pessimist may be correct in her pessimism that she's not likely to succeed, and by believing this, she makes it her reality.

CONTROL FREAKS

While we all seek control, some of us *seek* control, *needing* to be in the driver's seat at all times, both in and out of the car. They need to be calling all the shots. To some, these are known as control freaks. To others, they are known as pains in the butt.

Dan Nainan is an Indian-Japanese environmentalist techno-geek turned standup comedian. (He's got a funny act, if you want to catch him on YouTube.) In addition to doing comedy, Nainan is a self-proclaimed control freak. Others have proclaimed the same. "I'm a professional comedian who tours the world doing clean comedy—fifteen countries last year alone," said Nainan. "More than a few [of] my clients have given me the feedback that I'm a control freak."[22] His need for control grew from the challenges he has faced in his work life. Before his career in comedy, Nainan worked for Intel Corporation, doing demos on stage. I can't say I've done this myself, and the inner lives of computers are something of a mystery to me, but I'm sure this would be a high-pressure situation with a low tolerance for chaos. In conditions like this, staying in control makes sense. "I basically had to be a control freak because I was doing extremely high-wire demonstrations of beta versions of technology on stage with [Intel's cofounder] Andy Grove," said Nainan. "I would always have a backup computer ready to go in case the primary failed." You don't want to be rebooting while you're in the middle of a demonstration on stage in front of an audience, with your boss waiting.

When he turned to a life of comedy, appearing for the likes of Donald Trump and Hillary Clinton, Nainan found that staying in control through his attention to details still makes sense. Nainan looks cool and calm on stage, very natural; but there's a lot of thought and preparation that goes into looking so loose and casual. That cool, calm, and easy-going manner takes a lot of preparation and planning. "For example, if I perform during dinner, people can't laugh with food in their mouths, and my clients get upset when I insist I will not perform during dinner," said Nainan. "Or, I tell the people I absolutely have to have a wired, not wireless, microphone. They always complain, but I insist on this because I've had many wireless microphones cut out during my performances."

Nainan stays firmly in control offstage as well. "I refuse to get into someone else's car because half the time they're on the phone or texting or whatever and I don't want to expose myself to the danger. It sounds strange, but to me it's hilarious that so many people are afraid to fly but they think nothing of driving, which is literally tens of thousands of times more dangerous. People actually applaud when a plane lands—when they get out of their cars, they should get a standing ovation!" People might think Nainan is a little crazy, but after reading the statistics on this stuff, I wonder if *we* aren't the crazy ones.

Seeking an unusually high level of control can have a downside though, particularly for those at the extreme end of the control-seeking spectrum of personalities. Some control seekers might seem quite confident on the outside while churning with fear and insecurity on the inside. By seeking control in every little thing around them, they are attempting to reduce their anxiety, even if the things they are controlling are not terribly significant.

Speaker, author, and consultant Edie Raether spent many years as a practicing psychotherapist. She often helped people in her practice who were in turmoil on the inside while on the outside, they were doing their best to clamp down and control the people around them.[23] So what, then, is Raether's view of the more extreme control freaks and what drives them? "It's an inside job," said Raether, describing how these people grow up in chaotic environments and go through the rest of their lives with a great deal of anxiety about the world around them. Some faced alcoholic parents, absent parents, or abuse, and had to grow up quickly and take care of themselves (and possibly others) in unpredictable environments. The insecurity with which they grew up often stays with them as they get older, even if they aren't aware of it since it's all they've ever known. Rather than changing their own inner world, they do their best to change the whole world around them.

"When you don't have control over your inside world, you tranquilize and sedate yourself by controlling the outside world and other people," said Raether. "It's a way to counteract their internal turmoil and their state of anxiety. If I can control the time, place, and energy of a spouse or others around me, it gives me a sense that I can predict my world, and it's a safe place."

Unfortunately for the control freak, not everybody likes being controlled. "People who control eventually will push other people away because they don't want to live with them or work with them. As others push away, their anxiety increases. It's a big problem unless through therapy they understand how they can change from the inside out."[24]

Raether has personal experience of control freaks as well. "My ex-husband is a perfect example. He grew up with his mother who was a recluse," said Raether, describing his difficult upbringing. "If I were to put a spoon down, he would run from the other room and put a plate on the counter so I would put the spoon [on] it. He had a problem if I put my skirt on first and then my nylons. There's a rigidity in controlling people. You don't color outside their lines. All of that structure tranquilizes them. When we would drive, he would honk at intersections in case someone was coming. They live with a doom-and-gloom attitude about what's going to happen next, and then this makes bad things happen. It just goes around and around and around." However, rather than reducing the risks they feel, their ongoing quest for control seems to perpetuate and accentuate the risks they face, placing their goal ever farther out of reach.

Raether and her ex-husband were married for three years, but she knew very quickly, within the first few months, that she was in trouble. After their marriage ended, he rapidly bounced on to his next wife and probably repeated the control-freak pattern. "I was his third wife, and he quickly went on to his fourth. He just grabbed on to the next person he could control." Raether's first husband had his own set of issues, stemming once again from his need for structure and control. The rules the control seekers create are not just a problem for those around them but also for themselves. "They have to live by their own rules as well. When they get drunk, they feel a sense of freedom and liberation. [One of my ex-husbands] was an alcoholic, but he's a nice guy and [we're] good friends. . . . He's another left-brainer like my third husband, but when he would drink, he was the life of the party, with all of that fun and freedom he felt when he was released. When it wears off, they're back to their velvet prison, their restricted mind state."

Raether herself seems to be a real free spirit, while her husbands she told me about were far from it. It seems like an odd fit at first, but along the way, while talking to Raether about her marriages, it all started to fit together, patterns emerging. We are all seeking something in each other, each of us dealing with the risks of the past and bringing it forward into the present and into our lives with each other, for better and for worse. It puts some real meaning into the old cliché that "opposites attract."

"They were attracted to me and loved my free-spirited nature," said Raether, reflecting on her life and the lives of many others. Her husbands were attracted to her, and probably she to them, not despite their differences but because of them. "We have this shadow self or hidden side that needs to come out. We choose people who seem to have what we lack to try to make ourselves whole." For someone who has anxiety and is seeking control, he might be drawn to a free spirit as someone who has the thing he lacks, the person who can heal the fears, risks, and need for control he has carried with him. "But once they're with a person like that, a free spirit, the free spirit is hard to control. You have something they want, and they get angry and don't understand why."

We seek control to reduce our sense of risk, but what do you do when the quest for control is itself a risk? There is a way out of the loop that is found by looking inward. It's not an easy trip, but it's probably worth taking.

BLACK CATS AND STATIC

Because of our inner need for control, we go to great lengths to exert control over the world around us and reduce our sense of risk. Failing at that, we'll go to even greater lengths to create the illusion of control.

Superstitions are one way to feel like you're in control even if you aren't. They're not all about black cats and walking under ladders, though. While we often seem to look down on superstitions as inferior and silly, superstitions are found everywhere around the globe and have probably been around since the dawn of man. We might think of them as inferior to science and rational thought, but in way superstitions are related to science as tools for trying to understand the interconnectedness of our world. From there comes knowledge, and knowledge is power, power is control, and control makes things feel okay. So it feels, anyway.

Superstitions are the result of the constant efforts of our minds as pattern-recognition engines, searching for all manner of patterns in the world around us. We're pretty good at it. We can recognize writing when it's greatly distorted or obscured, as with the websites that require you to read and enter text to prove you're not a web robot. Our pattern-recognition engine leads us to see human faces in toast, car grills, and emoticons because our pattern-recognition engine is primed to find this pattern everywhere we look.

We also scan our world to find patterns in the events around us. If you walk under a ladder one day and then fall in a hole and break your leg the next, then your mind might automatically connect the two events without working through the details of how this might work. When the black cat walks across your path and you later slip and twist your ankle, your mind might put the two together, silently cataloging the simple belief that black cats cause twisted ankles. We might not have any idea why or how these two things might be related, but if we see this pattern showing up repeatedly, our mind might cement the apparent relationship in place.

While our minds are good at seeing patterns and holding onto them, we're not always as good at understanding how or why they are related. But our failure to understand the connection doesn't mean events aren't related. Both science and superstition are a consequence of our pattern-recognition engine at work. The difference is that science is a formalized system of harnessing the pattern recognition and putting it to work, compared to superstition, which dispenses with the formal part, operating all on its own.

Animals don't do science, for the most part, but they do learn to connect events in a way that looks like superstition. Pavlov's dogs didn't know why the bell was ringing whenever they got fed; they only knew that the two seemed to be linked. The learned response of the dogs looks like a doggy superstition. If you were an early human and you got attacked by a panther while going down to the watering hole on a night when the full moon was out, you might form a superstition that the moon is bad luck. If you're a modern human and you won a big baseball game when you were wearing your lucky underwear, you might decide to wear that underwear for every game. Hopefully washing the underwear doesn't wash the good luck out.

People don't like to admit to them, but superstitions remain surprisingly common. Buildings without a thirteenth floor, lucky shirts, and lucky lotto numbers are everywhere. The fact that superstitions are so widespread in many cultures, and still so common today, suggests that there is more to the formation of superstitions than old-fashioned, flawed thinking. We seem wired for this way of managing risk.

Researcher Kevin Foster at Oxford University thinks that the human tendency toward superstition might be the result of evolution, helping us to make sense of the world and feel like we can exert some modicum of control of it, even if the sense of control is often an illusion. Foster and his colleague Hanna Kokko built computer models to describe these weak associations between events that are involved in superstitions and to determine how they might affect evolution. In their models, associations were made between events that may or may not have been connected to each other, just like our minds are constantly doing behind the scenes as they look for patterns. Since these associations were random, many of them linked events that really had no connection. The computer weighed the benefits or costs of these associations, assuming that even if the superstitions were wrong and did not help the compu-organism much, they did not carry a great cost. Occasionally the computer, like our minds, made associations that were correct and actually did matter, helping the modeled compu-organism to survive. Under these conditions, with the pattern-recognition engine constantly spinning away in the minds of our ancestors, the tendency to create superstitions becomes an inevitable consequence of evolution because it helps creatures to survive. It should come as no surprise then that superstitions seem as common in people today as they ever were.[25]

After concluding their study, Foster and Kokko wrote about how superstitions may have evolved, even if most superstitions lack substance. "From here, the evolutionary rationale for superstition is clear: natural selection will favour strategies that make many incorrect causal associations in order to establish those that are essential for survival and reproduction."[26]

In nature and in the lives of early humans, superstitions may prove useful because sooner or later, in addition to those superstitions that lack substance, some of them may actually prove to be true. You have to avoid a predator and stay alive only once for a superstition to prove valuable. And usually the wrong superstitions don't cost us a great deal. Personally, I don't think black cats are actually bad luck, but if I did and I stayed away from them, I probably won't be harmed because I did this. "This argues that superstitions are the adaptive outcome of a general 'belief engine,' which evolved to both reduce anxiety and enable humans to make causal associations," wrote Foster and Kokka.[27]

Maybe the moon was providing light for the predators waiting at the watering hole, making it easier for them to strike. Maybe walking under a ladder is likely to end up with something falling on your head. Maybe I'll look out for those black cats next time I'm out.

YOU'VE GOT TO KNOW WHEN TO HOLD 'EM

The less control we have, the more prone we are to resort to superstitions to provide at least the illusion of control in risky and uncertain activities, like gambling, for example. The house always wins, they say, and statistics say the same thing, but gamblers don't always see it that way. People constantly hold out hope that while others may lose, they will be the ones who win. For some people, though, gambling can get out of control, becoming not a hope or a diversion but an obsessive addiction in which people gamble away large sums of money with the fundamental belief that one day they will finally beat the odds.

To get a look at the thought processes of gamblers, Prof. Robert Ladouceur from Laval University in Quebec set up gamblers at a video gaming terminal, gave them credits worth ten dollars, and told them to talk aloud freely while they played, saying whatever came to their minds. And they did just that, and all the while, the researchers listened in.[28]

Psychologists talked to the players before they started gambling and sorted them into two groups, labeling them as pathological gamblers or nonpathological gamblers. The observers found that while players in both groups talked while they gambled, they weren't saying the same things. The pathological gamblers made "erroneous statements," saying things about gambling that were not necessarily true and that overemphasized their odds of winning or having control over the game. They would talk to themselves, saying things like "If I touch my hat, I win more." Or, after losing several games, they would say, "I'm due for a win." Another term for these erroneous statements is *superstitions*. In addition to looking at what the players said, the observers looked at how strong their "convictions" were in the false belief. The more the pathological gamblers played, the greater their convictions in their false beliefs became, even as they lost. For the nonpathological players, their convictions in their false beliefs went away over time. Experience proved too strong for their false beliefs to persist.

Pathological gamblers are more superstitious than nonproblem gamblers, with superstition boosting their sense of control and making them feel less threatened. The problem is that they lack control and, through their beliefs, believe they are regaining control even as they lose their money. When the pathological gambler wins at least some money, as they are bound to do, they attribute it to their own control while the nonpathological gambler just chalks it up to luck.

OLYMPIC GOLD

Athletes, who also deal with intense pressure and uncertainty, are famous for their superstitions. Even the most popular of athletes, like Michael Jordan (who wore his old college shorts underneath his uniform during his entire professional basketball career, believing it helped him on the court) are not immune. Everyone knows it took more than a pair of shorts to make Jordan a spectacular player, but wearing the shorts probably did not hurt. In fact, some studies suggest that superstitions and the sense of control they provide might just help athletes.

Jennie Finch is an Olympic gold medalist in softball, world renowned as a leader in the sport, and recently retired. Finch cuts a striking figure both on and off the field—in 2005 she posed for the *Sports Illustrated* swimsuit edition. When she pitched to major-league baseball players, she frequently struck them out, catching them off guard with the speed of her pitches. To achieve this level of athletic performance requires superb physical conditioning and years of training. Many of the world's top athletes, including Finch, have found that their mental state is also important if they want to achieve peak performance, and they realize that superstitions may help them to reach this optimal mental state.

As part of her preparations for every game, Finch always followed a routine, starting with her uniform (as told to the author)[29]:

- Put my uniform on the same way, right foot in socks and pants first.
- My college roommate and I started spraying our uniforms with perfume before we left for the game.
- I always start warming up to pitch "27" minutes before game time. I wear #27.

- I like to put my bat bag and equipment in the same spot every game and between each inning.

Of course, the routine doesn't take care of everything. "Lastly, give it all I have and compete 'til the very last out!" Finch said.

You might think that an athlete like Finch would have been supremely confident when she walked onto the field, but, like anybody else, she felt the tension of the moment. The trick was finding a way to manage the tension to help and not hurt her on the field. "My butterflies were a good thing," said Finch. "I was always excited and, when it came down to it, if I prepared like I knew I should have, those butterflies were flying in formation. They were controlled."[30]

Going through her routine helped harness her fear and channel it, using it to her advantage. "I think my 'routine' helps me. My preparation is always the same. I think it also helps because it takes away some of the nerves when I am focused on my routine itself."

Where does a routine like this come from? I wondered. Why the number twenty-seven, for example? "My parents starting dating on the 27th their freshman year of high school," Finch said. "It's been their number since and a family number that I started wearing when I was 11 and until I retired."

Jennie's moved on from softball, spending time with her husband and family in Louisiana and raising two young children. Her husband, Casey Daigle, happens to be professional baseball player—a pitcher. And he has his own routine.

"He puts his uni on the same way every day, left side first," Finch said of her husband. "He only touches his game glove on game day. He puts his glove in the same spot when they are the home team. He warms up with the same routine, same amount of pitches and the way he finishes."[31]

We'll never know for sure how these routines affect athletes like Jennie Finch, but we can study how superstitions affect other people performing physical tasks. A group of students in Germany were asked to participate in a study in which they were putting golf balls.[32] Some students were given a ball they were told was a lucky ball, one that produced more successful putts, while others were given a ball that was labeled unlucky. The balls were all identical, but not to the students, who had the idea of the balls' luck planted in their minds. As a result of this idea, this superstition, the two groups of students performed differently. Those who felt they had a lucky ball managed to sink their putts almost 60 percent of the time, while those with a ball they felt was unlucky made only 40 percent of their putts. The superstition about lucky balls made a real difference in their games.

LUCKY MONEY

Like athletes, stock traders also deal with risky and uncertain conditions, often under great stress. Traders might be assumed to be cold and rational, using experience and analytics to get a leg up in the market, but their nerve-racking jobs also lead them toward superstitious behavior. "I think that for many of them the superstitions and the little rituals help them decrease the level of the pressure, the tension and the anxiety," said Dr. Bennett Leventhal at the University of Chicago.[33] Interestingly, while superstitions may help athletes, they seem to have the opposite effect among stock traders.

In a study examining trader superstition, a group of traders were placed in a computer trading-simulation exercise and were told that pressing X, Y, or Z on their keyboards might influence their results.[34] In truth, pressing the buttons had no impact, but, not knowing this, many of the traders tried pressing them. Those who stuck with the superstition, pressing away on those buttons, performed poorly in the simulation . . . but that's not the end of it. Those who relied on superstition in the simulation also had lower performance in their real-life work. While their anxiety might have been reduced for a brief time, overreliance on superstition may have kept them from dealing with real risks more effectively.

Superstitions come in many forms, and in China, numbers are often associated with good luck, particularly the number eight. Sometimes investors will put money into companies that happen to have a code for their stock containing the number eight; the more eights, the better, no matter the underlying financial fundamentals. If enough people share the belief in these investments by number, the superstition can pay off. Investments with the number eight involved do indeed tend to do well—the more people who believe and invest together, the more they will all be correct, making the superstition a winning bet.[35] With that in mind, I'm thinking of starting the 88888888 Company.

CONSPIRACY THEORY

One reason that superstitions are so universal is that they make us feel like we are in control even when we aren't. But this goes way beyond black cats. Our quest for control can change how we see the world, even leading us to see things that aren't really there.

Researchers Adam Galinsky and Jennifer Whitson of the University of Texas–Austin did a study in which they made some of their subjects feel that they lacked control. Then Galinsky and Whitson examined how well the subjects could see patterns. To make people feel like they lacked control, researchers asked them to remember a particular situation in their past in which they had lacked control, and the memory returned the subjects to that frame of mind. Another way to create the feeling of lacking control was to put subjects through a computer task where they had to guess a pattern of symbols and then the researchers would give the subjects random feedback, making them feel confused and helpless. After the feeling of lacking control was induced, all of those in the study were asked to identify simple shapes, such as a star or a ship, on a screen hidden in snowy static.[36]

The trick in the study was that not all of the images they were shown actually contained a shape; some of the images were pure static. Both groups did well at finding the hidden image where there was one to be found, but the group that felt out of control also went so far as to find hidden images when nothing was there. Their pattern-recognition engine was working overtime to reclaim their sense of control.[37]

This sense of seeking control through pattern recognition also extends to our propensity for conspiracy theories. Like superstitions, conspiracy theories abound, from the Masons to the JFK assassination, stitching together unrelated facts to try to create a coherent story. There's Area 51 (watch *The Roswell Incident*, or maybe *Independence Day*), Opus Dei (read and watch *The Da Vinci Code*), the Trilateral Commission (which rules the world), the moon landings (which were faked), and the attacks on Pearl Harbor (FDR knew it was coming). In one conspiracy theory, the headquarters for the New World Order is supposedly located beneath Denver International Airport, with Blackhawk helicopters ready to take over when civilization breaks down. (But keep that one between you and me—loose lips sink ships.)

Conspiracies don't usually need to have consistency or solid foundations in facts to support their cases, but still they are popular and persistent. Galinsky and Whitson had subjects in another part of their study think of a situation in their lives in which they either had been strongly in control or did not have any control. Next, researchers had the subjects read stories about people and then try to judge if conspiracies were involved in the stories.[38] In one scenario the participants read, they were up for hypothetical promotions. The day before a meeting with his or her boss, the subject notices a coworker e-mailing their boss and, afterward, finds out that he or she did not get the promotion. Were the coworker and their boss conspiring to block the promotion, or were the e-mails just a coincidence?

The people who remembered an event where they lacked control were likely to believe that a conspiracy was afoot in the story. Or, if the character in the story was not successful, they blamed a conspiracy theory that prevented him or her from succeeding. "Lacking control without an opportunity to self-affirm led participants to see images that did not exist and to perceive conspiracies," wrote Galinsky and Whitson.[39]

As strange as it might seem, seeing these patterns in static or believing in conspiracy theories might not be altogether bad. Developed as tools for dealing with our uncertain world, superstitions and conspiracy theories may lead us astray, but, by helping us feel more in control, they may also serve a useful purpose at times. "Illusory pattern perception may not be entirely maladaptive," Galinsky and Whitson wrote. "If pattern perception helps an individual regain a sense of control, the very act of perceiving a pattern, even an illusory one, may be enough to soothe this aversive state, decreasing depression and learned helplessness, creating confidence, and increasing agency."[40] If it makes us feel better and doesn't hurt anyone, then what the heck, I guess. We'll never be able to control everything in our world, but at least we can control ourselves.

CHAPTER 7

THE STORY OF VACCINES AND AUTISM

Why Stories Rule and Statistics Don't

HERO AND VILLAIN

Dr. Andrew Wakefield has had a big impact over the course of his medical career, though probably not in the way he expected. Growing up in the United Kingdom, both of his parents were doctors, and he studied medicine as well. After studying as a surgeon and working on transplants, he was introduced to a potential connection between autism and gastrointestinal disorders; from there, he grew a story that galvanized the world into two camps. On one side is the autism-activist community that fervently believes in Wakefield as a hero who stood up for his beliefs, despite all manner of grief the world hurled at him. On the other side, the medical community has virtually ostracized him as a fraud and a danger to the health of children. And there in the middle stands Wakefield, this highly polarizing figure, a hero to many and a villain to others, looking at first glance like an ordinary, good-natured fellow.

Autism affects millions of children and seems to be increasing rapidly, but its causes are poorly understood. Children develop autism in the first few years of life, causing them to regress socially, retreating from the world. Varying greatly, autism causes more severely affected children to lose all verbal and social connections with the world, even with their parents. Children can fail to develop normal speech, fail to make eye contact, and respond unusually to touch. Autism diagnoses seem to be increasing at an amazing rate, with any number of explanations thrown at the problem to see if anything sticks. Studies suggest autism is related to genetics and also to environmental causes, perhaps including exposure to pesticides or other toxins. Nobody really knows for sure, though. It can be heartbreaking for parents to see their children go down the path of autism and withdraw from the world, with no clue as to what is causing it and feeling powerless to stop it.

Wakefield believed that the gastrointestinal distress seen in some autism patients might be an important clue. He reasoned that the gastrointestinal problems might be caused by the MMR (measles, mumps, and rubella) vaccine that is given at an early age to virtually all children. The virus in the measles portion of the vaccine might infect the intestinal wall, he thought, making it unusually leaky and, in addition to causing GI problems, might allow unusual proteins to enter the brain and trigger autism.[1]

To look further into this autism-gut connection, Wakefield did a study on twelve children, which was published in 1998 in the prestigious British medical journal the *Lancet* with the pedestrian title "Ileal-Lymphoid-Nodular Hyperplasia, Non-specific Colitis, and Pervasive Developmental Disorder in Children."[2] The children in the study had been observed to have developmental disorders related to autism, and all twelve children in the study were said to have unusual inflammation in their intestines, which the paper suggested might be linked to autism. The authors suggested further study into the potential role of the MMR vaccine in causing gastrointestinal inflammation, although it did not go so far as to say that the vaccine caused autism.

The paper alone might not have stirred a great deal of excitement, certainly not based on its title, but Dr. Wakefield went a step further and held a press conference announcing a link he felt existed among the MMR vaccine, gastrointestinal illness, and autism. He claimed that eight of the twelve children had started observing developmental problems within days of getting the combined MMR vaccine. "I can't support the continued use of these three vaccines given in combination until this issue has been resolved," said Wakefield.[3] That was all it took. A story was born, a spark igniting the fears of concerned parents, and once it was started, the story took off, growing into a wildfire and consuming all involved.

Wakefield's theory seemed to put together all the pieces for parents worried about autism, helping the story to gain momentum. Babies normally get the first dose of the MMR vaccine around the age of twelve to fifteen months, and the first signs of autism are often seen in the first two years of life, so it seemed reasonable to link the two. The impact of the study grew into a worldwide scare, with anxious parents keeping their kids unvaccinated out of fear. Measles is a serious illness that can lead to death in some kids who contract it. Unfortunately, the rate of vaccination in the United States and in the United Kingdom plummeted, and the incidence of measles has increased for the first time in decades, all traced back to Wakefield's study with those twelve children. The fear of the MMR vaccine has led some parents to distrust other unrelated vaccines, like the pertussis vaccine, which has led to an increase in whooping cough, another potentially fatal illness for infants. By avoiding the perceived risk of autism because of Wakefield's story, some parents are exposing their kids to real and widely known risks of contracting these diseases—this is bad news for public-health officials who work to keep things like this from happening.[4]

Since Wakefield's press release, the story connecting autism and vaccines has collapsed under a growing burden of proof against it. Repeated large studies with thousands of kids have failed to find any link between autism and vaccines. In February 2010, the original paper was retracted by the *Lancet* after it was found that Wakefield apparently fudged the results, perhaps influenced by money he had received while doing the studies. British journalist Brian Deer found that Wakefield had altered or misrepresented medical records that were supposed to show that autism symptoms had appeared just after vaccination. Wakefield has continued to claim his innocence and has continued to support his initial hypothesis, but he lost his medical license in the United Kingdom.

Still, even with all of this, nothing will change the minds of scared parents whose fear pushes them to ignore all statistics or scientific information. Although Wakefield has been discredited and cast out from the medical community, from his new home in Texas he continues to be broadly supported and cheered by many parents and autism activists. In the face of a parent's fear and anger for his or her child's future, the story and the hope it provides is just too powerful for many people, against which statistics from studies have little influence.

THE STORY OF STORIES

Statistics can be a powerful tool to judge risks rationally. Insurance companies use statistics to measure risk for a reason—because these companies can't afford to trust their "feelings" when they have so much money on the line. Whether we're talking about the chances of a shark attack, a car accident, or a meteor hitting you, every risk has a probability that it will happen to us based on what has happened to other people in the past. An insurance company can't tell you that you will have a car accident tomorrow, but if they look at, say, all thirty-five-year-old females with a clean driving record who drive Honda Civics in Cincinnati, they can get a pretty good idea of the odds that it will happen. If you compile enough information about people and the bad things that happen to them, you can make useful assumptions about how likely it is that bad things may happen to you, and you may even be able to avoid them.

The problem with statistics is that we don't see the world statistically. Here, yet again, our nature works against us. Humans evolved to understand the world through a combination of stories and direct experience; a single powerful story or personal experience is usually far more than enough to overwhelm even the most accurate statistics. Statistics may lie, but stories can lie even better, and the same goes for telling the truth. The feelings that good stories generate can connect powerfully with how we see risks, changing what we see, say, think, and do. Statistics are great analytical tools, but they often fail to connect with our emotions. When this happens, they fall flat and have little impact.

One sign of the power of stories is that they are everywhere around the world, often with very common themes, and it seems like they always have been. The media used for stories may have changed from clay tablets and oral traditions to podcasts and 3-D movies, but we always seem to come back to the same basic stories. We tell stories about the creation of the world, the end of the world, how we should live and die, love and family, how nature works, and why we should avoid wolves dressed like our grandmothers. The act of telling a story involves the creation of a "virtual world" apart from the one in which we sit. Understanding a virtual world like this, separate from the one we live in, is a particularly challenging task for the brain, and odds are that whales, chimps, and other animals don't tell stories the same way we do, with the wolves telling stories about Red Riding Wolf. As an activity to which we devote a great deal of time, the origins and power of storytelling may well lie embedded in our evolutionary history.

Belonging to a group is crucial for us (see ch. 8), and good stories help to bind us together in society, giving us a common framework to understand ourselves and each other. They deal with the fundamentals of life like finding a mate, triumphing over adversity, surviving, and being part of the group—in other words, all of our deepest drives, dreams, and fears. By hitting us on a deeply emotional level, stories have a powerful impact on how we deal with risks. When you listen to a story and are drawn into that world, many of the same neurons in your mind start firing as if you were experiencing the events yourself. A good story feels real, whether you are watching it, reading it, or listening to it around a campfire.

"Storytelling is one of the mechanisms that play an important role in social bonding of communities," said Robin Dunbar, professor of evolutionary psychology at Oxford University. "Sharing the same worldview and culture seems to be an important key to facilitating cooperation and altruism on a wider scale. In one sense, that is indeed one of the hallmarks of being human."[5]

Evidence of the importance of stories in our lives can be seen in the plots of the greatest of human stories: those of our survival. These tales follow the rules of behavioral ecology, with evolutionary pressure seeming to align morality with biology. Natural selection is all about how well genes pass from one generation to the next. Anything that helps your genes to move on to the next generation might be thought of as a good thing, increasing your fitness. Anything that reduces the ability of your genes to move on to the next generation decreases your fitness and seems to be a bad thing.

For example, stories tell us to protect our families and children because it is morally correct, and, as it happens, protecting our families also increases our evolutionary fitness. Therefore, some of the oldest plotlines may have their basis in biology.

Most stories tell us that you don't kill your relatives or betray your own social group; a hero defends his family, his society, and his way of life, even to the death, while the villain is guilty of betrayal. The worst villains betray their own families, the most heinous of offenses. Mordred betrayed his father, King Arthur. King Lear was betrayed by his villainous daughters. Gaius Baltar in *Battlestar Galactica* betrayed all humanity and almost helped annihilate the human race. Villainous activity like this decreases your evolutionary fitness, since your relatives share some of the same genes found in your own blood and gametes. Contributing to the destruction of your whole species would result in very poor evolutionary fitness indeed.

Darth Vader of *Star Wars* was about as villainous as they come and would have been guilty of very poor fitness if he had killed his only children, Luke and Leia. Like many good villains, he was not purely evil in the end, having had a change of heart, thank goodness. When Vader saved Luke in the end, protecting him from the evil Emperor, he was redeemed and, in turn, also helped increase his evolutionary fitness by saving his son.

Defending your family and acting as the hero, on the other hand, is not only good for your image but also improves your evolutionary fitness, even if you sacrifice yourself, through what is known as *kin selection*. Kin selection explains some aspects of evolution that at first seem hard to understand. "If survival of the fittest is all about passing on our genes to the next generation, then how do you explain altruistic behavior, sacrificing yourself for others?" you might ask. At first, sacrificing yourself for others might seem to run against natural selection because your genes would seem to die with you—but that's not necessarily true. By helping your relatives, you're really helping your own genes as well because you are related and share genes with those relatives. Social insects like ants or bees may sacrifice themselves, but since they're all related in the hive or anthill, this sacrifice makes evolutionary sense, helping to pass on their own genes indirectly. Ground squirrels are more likely to help relatives—rather than nonrelatives—by using an alarm call and risking the attention of predators. Humans around the world are more likely to help relatives than nonrelatives.

The stories of good versus evil and hero versus villain can be complicated, but perhaps the roots of good stories are enmeshed in simple urges stemming from our evolutionary past.

HOW TO MAKE BAD DECISIONS

One of the problems with using statistics to judge risks is that we don't really understand what the statistics mean. This isn't all that surprising; statistics are often counterintuitive. I remember frequently feeling lost in my statistics class, and I know I wasn't the only one in the lecture hall doodling instead of listening. We sometimes think we understand statistics by going with our gut instinct, and we end up getting things completely wrong. If I tell you that I just flipped a coin and got heads ten times in a row, what are the odds that the coin will be heads on the next flip? Your gut may feel that the previous flips somehow alter the situation, even if your head knows that the answer is still 50 percent, no matter how many flips have come before.

Even experts like doctors have trouble using statistics to judge risks. German psychologist Gerd Gigerenzer established the Harding Center for Risk Literacy at the Max Planck Institute for Human

Development in order to improve how people use statistics in decision making. In one line of research, Gigerenzer has tested doctors for their ability to interpret a woman's risk of breast cancer, given a positive mammogram test. When they're tested for their ability to interpret these tests, doctors were told that 1 percent of women given a mammogram have breast cancer and that the mammogram finds this cancer 80 percent of the time. Also, mammography produces a false positive result in 9.6 percent of women who do not have breast cancer. The doctors are then asked, "What are the odds that a woman with a positive mammogram has breast cancer?"[6]

In one test of 150 gynecologists, only 21 percent got the correct answer. In a group of German doctors, the most common answer doctors gave was that 90 percent of women with a positive mammogram actually have breast cancer, with individual doctors saying anything from 1 to 90 percent. What's the correct answer? I'll give you a hint: 90 percent is not correct—it's not even close. Remember, this problem was given to doctors experienced with mammography, not random people off the street. The fact that so many doctors were unable to give the correct answer makes this a concern for any of us who might see a doctor at some point in our lives.

After the initial test, Gigerenzer and his colleagues teach doctors a different way to think about mammogram results using what he terms *natural frequencies*, describing a typical group of one thousand women given the test. Here's how it works: in one thousand women, ten will have breast cancer (1 percent of the group), and the test will detect cancer in eight of them (80 percent). For the 990 women who do not have cancer, about 9.6 percent will also have a positive test, because of the false-positive rate in mammography, meaning that ninety-five women or so will have a positive test even though they don't have cancer. Out of the one thousand women tested, this means there will be 103 with a positive test (95 + 8), out of which, eight actually have cancer. For women who have a positive mammogram, eight out of 103 actually have cancer, about 8 percent—far fewer than 90 percent. While most people (including doctors) think a positive mammogram means they're almost certain to have cancer, most people with a positive mammogram do not have cancer at all.

Our failure to understand statistics about risk creates dangers of its own. If doctors don't understand know how to interpret a positive mammogram, they won't correctly communicate the results to patients. For the women who have a positive mammogram, many will assume the test means they definitely have cancer, and they develop intense dread and fear of the C-word, assuming it's a virtual death sentence. "What we call the 'illusion of certainty' is widespread," said Markus Feufel, a researcher at the Harding Center for Risk Literacy.[7] Some patients who have received positive mammograms even have procedures performed that are unnecessary and perhaps harmful. "Mammography screening also detects so-called 'indolent' (slowly growing or less aggressive) tumors, which would never develop into a life-threatening disease. But because their development cannot be predicted, about 10 of 2,000 women in the screening group had their breast completely or partially removed, though this would not have been necessary," declared a summary about mammography from the Harding Center.[8]

Breast cancer is not the only medical condition where we have trouble understanding risk. As men age, they are commonly given a diagnostic test for prostate cancer that measures the levels of a protein called prostate-specific antigen (PSA) in their blood. In many patients who have prostate cancer, their PSA levels rise, and PSA can easily be measured using blood samples without anyone needing to bend over. The problem with prostate cancer screening using the PSA test is that it also yields many false positives, just like mammography. According to the Harding Center, for every one thousand men screened for prostate cancer with this test, 180 will have false positive results and not have cancer at all. The biopsy used to look further for prostate cancer involves multiple needle punctures of the prostate, usually through the rectum—not a fun thing and with risks of its own, such as infection. And for those one thousand men tested, ten to thirty of them will be diagnosed with cancer but treated unnecessarily with prostate removal, radiation therapy, or other procedures. And the net impact on prostate cancer deaths? The studies indicate that men who were not screened die of prostate cancer at the same rate as men who were screened.[9]

Maybe I should have paid more attention in statistics class.

THE INSURANCE SALESMAN'S TALE

As a State Farm Agent in San Diego, California, Matt Kalla sees risk from a different perspective than most of us.[10] Selling life, auto, home, and other insurance policies, he deals with losses ranging from death to fender benders, from massive wildfires to overflowing bathtubs. While most of us live with the basic mind-set that nothing bad will ever happen to us, Matt knows this isn't true. He sees bad things happen every day. "We're all going to die, but a lot of people act like they don't know it," Matt said when I spoke with him. "You're going to have a car accident someday, whether it's you or the other guy who causes it. You're going to get sick." The statistics say that sooner or later these things will happen to all of us. But it's something we manage to avoid thinking about as much as possible.

I've known Matt for several years. He's worked in the insurance industry for twenty years, taking over the business from his father a few years back, and he has an office in part of San Diego known as Point Loma, a peninsula along the bay and ocean. His office is in a simple building with a blue awning, between a hair salon and a veterinarian. He's a friendly, regular guy—nothing like the obnoxious insurance salesman in *Groundhog Day*, who pesters Bill Murray. From Matt's perspective, his job is to help people, to give them protection when bad things happen, because they are bound to happen whether or not we like to think about it. "Money changes lives," Matt said. "I see it when I hand someone a half-million-dollar check and see the weight lifted in their eyes." From the customer's perspective, however, insurance can seem like a nuisance before the bad thing happens, something they're spending money on and don't see any immediate return for. When money is tight, as it has been for many people in the last few years, people are reluctant to spend money on insurance.

First-year statistics on payouts from claims are one interesting set of facts Matt has on hand when he talks with clients, describing how many people have claims paid in the first year they are insured. Whether it's for car accidents, heart attacks, or anything else, these people signed up just in the nick of time. "A person has to feel like this could happen to them," Matt said, describing how important it is for clients to feel the real benefit they are getting for their money.

While customers often have an idea about the statistics, this isn't usually what changes their mind about insurance. "By far the best thing to do is to tell a story," said Matt. "And it's even better if the story is closer to home, if it's someone they know, like friends or family. These can be terrible, but [they are] the kind of stories that you try to bring to the table." A story that makes you think that, if you died yesterday, how much insurance you would have wanted today.

Matt remembers the story of a friend of his, a man, forty-seven years old, who was volunteering at a winery. Because of his allergic sensitivity to bees, the man carried an emergency autoinjector, like a pen, to inject epinephrine in the event he ever got stung. While at the winery one day, his epinephrine autoinjector was in his car but not in his pocket when he got stung and went into anaphylactic shock. "He never even made it to the car," Matt said. "He had a wife and two kids left with no life insurance, no job, and a mortgage to pay." The moral of the story is clear: a little life insurance goes a long way. And don't leave your EpiPen in your car.

Another story Matt remembers is about long-term care, insurance that provides for expenses for daily living when people are disabled and unable to take care of themselves. This isn't just for the elderly. Among those who need long-term care in the United States, 40 percent are under the age of sixty-five, and the group includes many people who have been injured in accidents. A State Farm agent Matt knew of had been thinking for years about signing up for long-term care, then he won a trip from the company to go to Hawaii. He figured that since he was selling the stuff, he should probably own it himself, and signed up for long-term care just before the trip. "In Hawaii, when he got there, he jumped in the water, broke his neck, and became a paraplegic," said Matt. "Because he got the policy, he's okay still and can run his office even." That's the kind of story that sells insurance, stories that deliver the emotional impact that statistics cannot. Stories that make us see and feel what someone went through and what we would feel like if it happened to us. A statistic might make us think "Really?" but is unlikely to hit us in the gut hard enough to get us to buy life insurance. "The basic starting point for most people is, 'it will never happen to me' until they hear a story they relate to," one that really hits home.[11]

THE SHOCKING EFFECT OF FEAR

When we are really afraid of something, scared out of our wits about it, all sense of perspective about how likely it is flees from our mind. The more we dread something, like a painful death from cancer or having a child develop autism, the worse we become at judging its risk. Scientists call this *probability neglect*, the ability of a disturbing image or event to make us blind to the probability that it will or will not happen. The greater the dread, the more awful the consequences and the less we care about numbers when judging the risk.

An electric shock is a distinctly unpleasant experience most of us dread, and it's a fairly dangerous one if enough current and voltage is involved. Asking volunteers (undergraduate students at Rice University) about a hypothetical study, Yuval Rottenstreich and Christopher K. Hsee (both researchers at the Center for Decision Research at the Graduate School of Business, University of Chicago) told them that, in the study, they would be subjected to a "short, painful, but not dangerous electric shock," then asked the students how much they would pay to avoid the shock if there was a 1 percent chance of being shocked, a 99 percent chance of getting shocked, or a 100 percent chance of getting shocked.[12] If there was a 100 percent chance of shock, students would pay $19.86 (the median response). To avoid a 99 percent chance of shock, students would pay ten dollars; and to avoid a 1 percent chance of shock, they would pay seven dollars. (Don't worry—nobody was actually shocked in this study. The most unpleasant thing they experienced was answering questions about hypothetical shocks. Other studies using real electric shocks have found the same results, though.)

One curious thing about the study is that, overall, the subjects were so worried about the idea of getting a shock (even though the shock was purely hypothetical) that a 1 percent chance of getting shocked felt about the same to those volunteers as a 99 percent chance of getting a shock. They were willing to pay approximately the same amount either way. There's more of a difference in their minds between a 99 percent and 100 percent chance of shock than there is between a 1 percent and 99 percent chance. Because the thought of getting shocked is so unpleasant, all that matters to the subjects, for the most part, is that there is any chance of getting shocked at all, no matter how large or how small the odds are.

"We perceive a substantial, qualitative difference between situations in which there is no risk and situations in which there is even the tiniest bit of risk," said Rottenstreich when we discussed the study.[13] "There is just something about the way in which we see, feel, experience the world (i.e., perceive things) that in essence makes us exaggerate small risks."

"A shorthand way to think of how peoples' perceptions of risk work is with three categories: no chance of something happening, some chance of it happening, and it'll surely happen," explained Rottenstreich. "When people think about probabilities like 0 percent, 1 percent and 50 percent, they seem to sense 0 percent as qualitatively different from 1 percent and 50 percent, and they might even sense 1 percent and 50 percent as relatively alike. Zero percent is in the no-chance category. One percent and 50 percent are both in the some chance category. So it's as if people are thinking categorically or qualitatively in the way just outlined rather than quantitatively. Because quantitatively it's clear that 0 percent and 1 percent are relatively alike and different from 50 percent."[14]

It's not just fear and dread that shape how we see risks. Positive emotions can also blind us to statistics. In another experiment, Rottenstreich and Hsee found that students paid no attention to the odds when they thought they might win an exciting $500 discount for a trip to Europe, but the students weighed the odds much more evenly when the outcome was a more mundane $500 discount on their tuition. Students apparently love the idea of going to Europe but are not as interested in reducing their tuition. (Perhaps another study should look at the responses of parents of college students, a neglected group in risk-perception studies.) And when the students were asked about kissing their favorite movie stars? Their feelings jumped back in and their rational analysis of probability jumped out the window.

Imbuing something with positive feelings is one way that marketing works its magic on us. Our rational mind might question if we really need a car or jewelry, and it will question the price being asked. Once our feelings come into play and push rationality out of the way, we see that car not as a boring mode of transportation but as something that will make us cool and sexy. We like being cool and sexy, making us less sensitive to numbers like costs. All we know is that we want it.

Probability neglect starts showing up all over the place once you start looking for it.[15] It's why we buy lottery tickets. It takes only a tiny chance of winning to make it seem worth a gamble to us; whether the odds are one in ten thousand or in one in ten million doesn't matter to us much. Probability neglect explains why travelers buy terrorism insurance for air travel at a much higher rate than all other causes, because terrorism is scary and gets your emotions going, blinding you to the low probability that this risk would actually be a threat to you. Studies also show that people who see scary movie clips or read disturbing newspaper articles view the whole world as riskier afterward; once their fear is stirred, it affects everything they see, whether or not the risk is related.

Probability neglect falls from the sky when you talk about meteors. If you look at the chances of getting killed by an asteroid or meteor, they are astronomically small, so small that we really don't know what the odds are exactly. The moon is littered with impact craters, but most of them date back billions of years. Once every ten thousand years you've got a good chance that a forty-meter asteroid will hit the Earth somewhere—that's about the size of the object that made Meteor Crater in Arizona fifty thousand years ago. And once every 100,000,000 million years or so, the Earth may see the likes of the type of impact that spelled doom for the dinosaurs, the type of impact shown in the movies.

The movies about meteor and asteroid impacts are spectacular. When you see the planet killer coming, and the astronauts and earthlings scrambling to do something about it (and usually succeeding, thank goodness), it's a real heart-stopper, an end-of-the-world scenario. All of this overwhelms your ability to look at the numbers involved. When you add it all up, the odds of getting hit by an asteroid are probably about one in a million, probably about a hundredfold lower than your chance of being struck by lightning. But you don't care about the odds because the thought of the end of the world is so emotionally charged that whether it's one in a hundred, one in a thousand, or one a million does not make much of a difference. We lose track of the zeros when so much is at stake.

PROSPECT THEORY

Economists like to build models that can predict how people will behave based on a variety of assumptions. One of the larger assumptions has always been that people balance the risks and rewards of our actions rationally based on the numbers involved.

For example, if two widgets are the same, but one widget costs three dollars and the other widget is four dollars, then the rational consumer would buy the three-dollar widget. The world of these models is often neat, clean, and predictable. Very nice. And also not very realistic much of the time, unfortunately.

The problem is that the real world is seldom so kind as to match the assumptions of economists. People are not calculators, and we are influenced by a great many things other than adding up columns of numbers in our head. The four-dollar widget might have better advertising, which makes it seem cooler and sexier to us, so we might buy it whether or not this is the rational thing to do.

Realizing this, a new breed of economists are trying to make models that fit the real world and its risks rather than trying to make the real world fit their models. One aspect of the real world explained by these models is our deep dislike of losing things. We are loss averse, to put it mildly. If I give you twenty dollars, for example, you will not value this as much as you would if I took away twenty dollars. Money that we lose is valued about twice as much overall as winning the same amount because we can't stand to let go once the cash is in our hands.[16]

Economist Daniel Kahneman of Princeton and his collaborator, Amos Tversky, and others developed what is known as *prospect theory*, which describes how we really act rather than how we should (earning Kahneman the Nobel Prize for Economics in 2002).[17] His work covered a broad range of topics in the budding field of behavioral economics, including how we fail to see risks statistically. Loss aversion is one aspect of this.

Prospect theory also underlies the observations of those like Rottenstreich and Hsee, that we value certainty far more than numbers alone would dictate, paying far more for a 0 percent chance of getting a shock than a 1 percent chance, for example. Many aspects of how we see risk feed into the workings of the economy, large and small, with psychology becoming an integral part of economics.

The more you look, the more you see that economics, like all endeavors, is all too human when you get into the real world. We fail to sense and prepare for rare financial risks, just like we undervalue other rare risks we haven't yet experienced, like earthquakes. We want control of our money, just like we want control of other parts of our lives. The more we understand ourselves and how we perceive risks, the better we will understand how we handle our money.

COMFORTABLY NUMB

When we hear the story of a single individual who has run into hard times, it stirs empathy in us (most of us, at least), and we want to help by giving our time, energy, and money to make a difference. It could be a story about a child with cancer, a tornado survivor, or a family left homeless by foreclosure. We connect with the person or people emotionally; we also connect with the story, giving it meaning for us and driving us to act.

A strange thing happens, though, as we hear about more and more people having bad things happen to them; rather than caring more, we care less. Research has revealed that we do this for those affected by natural disasters, illness, famine, poverty, or environmental problems like climate change. We also do this when talking about the millions of people killed around the world in genocides.

An estimated fifteen million to seventeen million people died in genocides during World War II. After the war, the world said that such atrocities would never again be allowed to happen. But they have— not just once, but repeatedly. Cambodia, Rwanda, Bosnia, India, Tibet, and Darfur are just a few of the post–World War II genocides that leave millions dead, with some genocides continuing for years with little done to stop them. Why don't we do more? Don't we care?

In a way, we don't.

The genocide in the Darfur region of Sudan is one example. It's hard to say exactly how many people have died in Darfur since 2003, when the genocide began, but it's likely that the dead number in the hundreds of thousands, with millions more forced to flee. "Darfur has shone a particularly harsh light on the failures to intervene in genocide. As Richard Just has observed, 'we are awash in information about Darfur. . . . No genocide has ever been so thoroughly documented while it was taking place . . . but the genocide continues. We document what we do not stop. The truth does not set anybody free. . . . How could we have known so much and done so little?'"[18]

While the death of a million would seem a million times as bad as the death of a single person, our minds don't work this way. We respond with an outpouring of kindness to the story of a single child left parentless after a car accident, but we walk away from the millions.

"On the one hand, we respond strongly to aid a single individual in need," wrote Paul Slovic, professor of psychology at the University of Oregon and president of Decision Research. "On the other hand, we often fail to prevent mass tragedies such as genocide or take appropriate measures to reduce potential losses from natural disasters. This might seem irrational but we think this occurs, in part, because as numbers get larger and larger, we become insensitive; numbers fail to trigger the emotion or feeling necessary to motivate action."[19]

The problem with the hundreds of thousands of people in Darfur who have been killed is that, with so many people involved, they change from a story to a statistic. Stories have emotional impact, hitting us in the gut and stirring us into action. Statistics don't. The reason for our lack of response to genocide or large-scale natural disasters is that our feelings don't scale up. We don't take the feelings we have about one person and make them one hundred thousand times as large when one hundred thousand people are involved. Our brain just doesn't work this way.

Charities know this, so they always put a name and a face on a tragedy. When Sally Struthers is on the television asking for money for impoverished children, she shows you the face of a specific child who needs your help because this is something to which you can relate and respond. It might not make sense, but charities know that naming one thousand people at risk, or millions who have died, will not produce results in the same way that a moving story of a single individual does.

Slovic and others call the way that we seem to value life less when it comes in large numbers *psychosocial numbing*. Because of this trait, the amount that we value life seems to decline as more people are affected, as if when large numbers are involved we no longer see them as people.

In one study to which Slovic contributed, people were asked to donate five dollars to the Save the Children fund. They were asked to do one of two things: they could help feed a seven-year-old African girl, named Rokia, and were shown her picture, thus creating a story; or they could help feed millions, a sheer statistic. The story with Rokia got twice the donations as the statistic of millions. And in a third group, where the story and the statistic were combined, the statistic actually hurt donations compared to the story with Rokia alone.[20]

"By understanding that our intuitions are not trustworthy in certain circumstances, then we have to work around them," said Paul Slovic when I spoke with him. "If we find that our vision is inadequate, then we wear glasses or do other things to compensate.

We need to use the appropriate compensations when we can't trust intuitions."[21]

"How should we value a human life? If we look at how we actually do value life, it's all over the map," said Slovic, "from a very high value for a single life that's identified, that we have experienced and know them personally, but incredibly low implicit value in the way that we neglect to protect large numbers of people."

Slovic proscribes some corrective lenses that might help us respond to large-scale suffering. "We should create laws and institutions to set those laws in motion. Laws and institutions are dedicated to thoughtful values and are persistent, not sensitive to the whims of the emotional moment." Laws and institutions may not be perfect, but they're probably better than not trying at all. "Laws and institutions are created by people, and the rules will be debated and fought over according to individual predispositions and prejudices. On the other hand, the alternative of going with what feels right is even more subject to those whims and [is] less transparent. At least with laws and institutions there's a public record to criticize and argue about. If you leave it to the intuitive whims of people, you don't know what's going on."[22]

While our gut instincts and emotional responses to stories may distort our views of risk, they can also help us sometimes. Giving statistics emotional value, creating a story around them, can make us better at using statistics in our decision making. In effect, they work with our biology rather than against it, by turning the statistics into a story.

It's all about relating to people as if they are real and something we care about. And this brings us to the risks stemming from our social nature, including the thing we fear more than death itself.

CHAPTER 8

THE THING WE FEAR MORE THAN DEATH

How Our Need to Belong Makes Us Vulnerable to Con Men and Fearful of Podiums

LONGING FOR BELONGING

Putting aside all of our differences, deep in our hearts, we are all fundamentally social creatures. We might not all be terribly sociable or extraverted, but we are still social one way or another. It goes right to the core of being human: "Existing evidence supports the hypothesis that the need to belong is a powerful, fundamental, and extremely pervasive motivation," psychology researchers Roy Baumeister and Mark Leery wrote.[1]

Perhaps if humans had evolved in a different environment, one without keen predation pressures, we might not have evolved such a compelling need for being social. Although we would probably not have been human then either. "In primates, group-living is a response to predation pressure," said evolutionary psychologist Robin Dunbar.[2] Given the importance that belonging to a group plays in our mind, it should not be surprising if we sense getting kicked out of the group as a great risk, maybe the greatest. We fear it like death because in the past being ostracized would often result in that very fate.

We tend to think of being social as a trivial activity, a waste of time with no clear purpose or value; but for most of humanity's history, being part of a social group meant survival. Before they were even fully human, living in groups helped hominids and other primates to fight off predators (see ch. 1). A group of monkeys (or hominids) can keep watch together, sound alarm calls for snakes, and attack predators as a mob, fighting off and even killing predators such as leopards.

If carried far enough, being rejected results in complete isolation, or ostracism. In the same way that we share our social nature with other creatures, humans are also not alone in our ability to ostracize other members of the group. "Ostracism appears to occur in all social animals that have been observed in nature," said Kip Williams, professor of psychological sciences at Purdue University and an expert in the science of ostracism and how it affects us.[3] "The usual reasons are because the targeted member of the group poses a threat or burden to the group. This could be because they are injured, because they behave erratically and unpredictably, because they are orphans and require too much care and attention, etc. To my knowledge, in the animal kingdom, ostracism is not only a form of social death, it also results in death. The animal is unable to protect itself against predators, cannot garner enough food, etc., and usually dies within a short period of time."[4]

While predators are no longer the biggest concern for most of us, our social groups have continued to grow in size and complexity. We have high-school cliques, workplace alliances, college sororities, gangs, and nation-states, all providing us with more or less the same thing—they help us to group together, belong to a group, and protect each other. I've heard that women even go together to the restroom. I don't think that men do this much, but maybe it's another way to express the same basic need to be social.

One of the greatest punishments practiced in some societies, including on playgrounds, is shunning. School kids often suffer rejection from the other kids, creating not just loneliness but great and lasting pain. The consequences of such ostracism can be dramatic. In an article he wrote about the science of ostracism, Kip Williams tells the story of a girl in kindergarten, Jennifer, who impulsively kissed a boy.[5] Afterward, the boy and his friends sprayed him with imaginary repellant to keep the girl away, and soon the other kids in the school were doing the same. "Within just a few days, none of the children would be caught talking with her or including her in their activities. Jennifer was no longer invited to parties, to overnights, to recess games. Jennifer was ostracized. Not for a day, a week, or even a month. Her classmates ostracized Jennifer until her parents decided to move her to another school . . . in fifth grade." I can only imagine the impact this had on the girl.

CYBER-REJECTION

One day, in the mid-1980s, Kip Williams was sitting in the park with his dog when a Frisbee rolled into him from a couple of nearby players. He stood to throw it back, and to his surprise they threw it back to him again, and so a game of catch emerged that went on for a few minutes, in what seemed to Williams to be a fun and spontaneous game. But soon they stopped including him, for no reason that he could see, simply going back to the original game with the two playing catch. Williams had never met the two Frisbee players before and had only briefly played catch with them. He had no reason to believe that there would be any relationship beyond a quick game of Frisbee, but even as a university professor at the time, he felt awkward and slightly hurt by the apparent rejection. He sat down and petted his dog to heal the own social wounds that even this small version of ostracism could cause. Later, pondering the power of such a small rejection, Williams started designing experiments to study ostracism.[6]

Williams and his associates created a computer game they called Cyberball to look at the impact of rejection on people. In this game, a test subject plays a virtual game of catch with two other players represented by simple cartoon characters on the computer screen. They believe the other characters are controlled by humans somewhere, although in reality they are controlled by the computer. Some of the subjects will have the cyberball tossed to them one-third of the time, as you would expect if all the players were equally included in the game. Other players will receive the ball one or two times and then be left out. They were cyber-ostracized.

With this lab version of ostracism under controlled conditions, it's easier to see how ostracism affects people than it would be in real life. One way to test this is by asking people who've played Cyberball to fill out a questionnaire afterward. When Williams and his colleagues did this, it revealed that even a brief ostracism with computer players in this simple game was enough to make people feel bad. "We found that those who had been cyber-ostracized for just a few minutes reported unusually low levels of belonging to groups or society, diminished self-esteem, and a lack of meaning in, and control over, their lives. They were also sad and angry," wrote Williams. All from a brief computer game.[7]

One Cyberball experiment gave players the choice to eat some yummy cookies or drink a less yummy health beverage after playing. The ostracized players ate more cookies and drank less of the health beverage, drowning their cyber-sorrows in not-so-cyber junk food.[8] The impact of ostracism can take a darker turn as well, leading to aggression. In school shootings, the shooters have often been kids who were cut off from the rest of the school and were lashing out at those whom they felt had rejected them. It's a sad and scary reminder that being social is an essential part of a normal, healthy human life, and the consequences of being left out can be great.

HERMITS AND SOCIOPATHS

Being social is probably a part of the basic definition of healthy human life, and failure to be part of a social group carries great risks. Those lacking in social interaction, such as hermits, have a mixed reputation. On the one hand, their image is as an unkempt, wild, and dangerous recluse, but on the other hand, they are sometimes highly spiritual and individualistic, pursuing their own courses. Whether you are one or the other depends on the details.

Henry David Thoreau spent time alone at Walden Pond and has been celebrated for it. "If a man does not keep pace with his companions, perhaps it is because he hears a different drummer. Let him step to the music which he hears, however measured or far away," Thoreau wrote, praising the virtues of pursuing your own course.[9] He lived alone in his cabin at Walden for two years, two months, and two days, and praised the solitary life, living self-reliantly. His cabin was not far from town, though, and was on the land of his friend Emerson, so he was not really much of a hermit. Emily Dickinson seldom left her home in her later years, but she wrote to friends and amassed a collection of more than 1,800 poems. She published few of these in her life, but still the act of writing 1,800 poems is a social one in a way, silently expressing her desire to communicate intimately what she must have had difficulty saying out loud.

There's a risk that the lack of human contact leads a person's thoughts to stray so far from the norm that he or she becomes a socio-path, thus becoming a risk to others when no longer restrained by the normal social order. This seems to be the case with Ted Kaczynski, the Unabomber. While Kaczynski, like Thoreau, seemed to prize self-reliance and a simple life alone in a cabin (even publishing a work cele-brating the value of a life free of technology and complexity), his life and work were distinct from Thoreau in several ways.

Kaczynski built and sent out mail bombs over a period of seven-teen years, killing three people and injuring many more, while Thoreau did not. But Kaczynski was not born a bomb builder, it seems. As a child, he was an early math prodigy, leaping ahead in school but playing little with other children, and graduating from high school at age fifteen. Getting a doctorate in advanced mathematics, he was hired on the faculty at the University of California–Berkeley at age twenty-four and stayed for two years. He was the youngest faculty member ever at Berkeley, but he resigned after numerous complaints about his strange teaching behavior. This seems to be a key point when Kaczynski started to withdraw, first to his family's basement and then moving even farther away from society to remote Montana.

It may not be possible to put your finger on what exactly caused Kaczynski to slip down the slope and away from the world. He may have been born with a predisposition that lead to a life as a social outcast, or maybe the brutal psychological experiment for which he had volunteered during his college years set him over the edge.[10] But somehow he slipped off the path of social connection and spiraled into a world of his own. From 1971 until his capture in 1996, the Unabomber lived alone in a tiny cabin in the Montana wilderness, trying to live off the land. As development encroached on his wil-derness, plowing a road through a favorite spot of his, he decided to fight back. Targeting university professors, the airline industry, and others he felt represented industrial development, he crafted home-made bombs to send in the mail. Over the years, he killed three people and injured many more, hoping that he could, all on his own, stem the tide of technology that was controlling our lives.

The FBI spent years searching unsuccessfully for the Unabomber, but what ultimately gave him away was his effort to reach out to others, his manifesto. Perhaps in his mind it was his version of *Walden*, hoping to spark a movement. Titled *Industrial Society and Its Future*, the thirty-five-thousand-word manifesto outlines the problems Kaczynski saw in modern, technological society, calling for a revolution to bring its end. Kaczynski wrote that people in our modern world had been forced "to behave in ways that are increasingly remote from the natural pattern of human behavior."[11] In his mind, the natural pattern was one with full autonomy for individuals, independent of society and social connections. One of the problems with liberals, he wrote, was that they were "over-socialized," in other words, they were too willing to play by the rules.

After publishing the manifesto, he was identified by his brother and soon captured. So why did he publish it? He had cut himself off from the world, but still he felt a need to communicate, to reach others. He had few ties with others by this time, including his own family, but writing the manifesto meant that, in his way, he had needed to connect, to make a difference in something bigger than himself, even if in our mind the difference was not a positive one. Even someone as estranged from society as Kaczynski felt the need to tell his story to others, to have an impact on the world, to connect somehow with the society that he worked so hard to be separate from. He was being social, in his antisocial way.

SOLITARY

Humans live in groups that cooperate and communicate to help each other, and not just because we want to but because we need to. Even the most introverted of us benefits from and yearns for human contact even as we agonize about it, and none of us is fully human without it.

This isn't to say people can't live alone. Some people live quite solitary lives for many years, eating, sleeping, and doing just about everything alone. For some, this is a religious commitment, living the life of a monk. But this is far from the normal human experience, and in most cases living alone over the long run leads to a suite of illness, both mental and physical. Maintaining basic physiological functions does not seem sufficient to say we're living fully.

While some people choose to live on their own, others have it decided for them, spending years in solitary confinement. Solitary confinement is commonly used in the United States to punish and segregate inmates, but it is viewed by many as an extreme form of punishment, like legalized torture, because of the severe impact it has. Nonetheless, its use has been growing. The United States houses twenty-five thousand criminals in solitary confinement, which is unique among developed countries in resorting to this measure to this extent.[12]

Having evolved as social creatures, the impact of depriving us of contact with others can be profound. Individuals held in solitary for long periods start to lose their sense of who they are, often exhibiting signs of psychosis, severe depression, and even insanity. For this reason it is often incorporated in torture protocols, hoping to cause an individual to lose his grip and loosen his tongue. Interestingly, being kept in solitary confinement often leads to psychosis in inmates who never displayed symptoms before.

"Some of them lose their grasp of their identity. Who we are, and how we function in the world around us, is very much nested in our relation to other people," said Craig Haney, social psychology professor from the University of California–Santa Cruz.[13] "Over a long period of time, solitary confinement undermines one's sense of self. It undermines your ability to register and regulate emotion. The appropriateness of what you're thinking and feeling is difficult to index, because we're so dependent on contact with others for that feedback. And for some people, it becomes a struggle to maintain sanity."[14]

One of the biggest concerns for long-range spaceflight is how astronauts will hold up in conditions with little social contact, when it could take hours for a reply from friends or family to travel through space, hundreds of millions of miles. It makes for an awkward conversation. HAL may not be the only one going bonkers on a trip to Jupiter. On missions to the International Space Station, American astronauts often show signs of depression, feeling isolated even with their two Russian comrades around. A trip to Mars or farther in space could take months or longer, pushing astronauts to the brink because of isolation, even if their technology functions flawlessly. Even back here on earth, during extended stays at research stations in Antarctica, crews can be isolated for many months while tensions rise and depression sets in over the long, dark winter.

THE PAIN OF REJECTION

Anything that blocks or threatens our connection with others feels like a grave risk, as serious as any physical injury. When you say that someone has "broken your heart," it's more than just a sentimental phrase in romance novels. When we are rejected by a group or when we are romantically rejected, it truly hurts, creating real pain and an increased risk of physical illness as well as making us hurt emotionally.

The pain of rejection is real enough that you can see it in the brain. People playing Kip Williams's Cyberball were put in an MRI to get a play-by-play view of what happens in our brain when we are left out. Curiously, the rejected people had their brain light up in areas associated with the emotional side of physical pain, explaining why rejection hurts so badly. And even more curiously, they found that if they gave Cyberball players Tylenol (acetaminophen) before playing, it blunted this response. It seems that Tylenol can block the pain of being rejected. "In a parallel experiment, the researchers also found that twice-daily doses of acetaminophen over three weeks reduced daily reports of distress and hurt feelings from social rejection in 62 students, compared with the effects of a placebo," wrote Williams.[15]

I wish I had known that back in school when we were picking teams in gym class.

The relationship between physical pain and the pain of being left out goes further. Many of the same pathways involved in physical pain are also involved in social pain. University of California–Los Angeles researcher Naomi Eisenberger found that the mutations in the gene that makes people feel pain relief with morphine also make people sensitive to social rejection.[16] The same neural pathways in the brain carry physical pain and the pain of being rejected, conserving ancient signaling mechanisms and adapting them to the social complexities of the human world. Feeling pain in this way ensures that we pay attention and try to do something about it in order to stay in the social group. It's that important.

"Because social connection is so important, feeling literally hurt by not having social connections may be an adaptive way to make sure we keep them," Eisenberger said.[17] "Over the course of evolution, the social attachment system, which ensures social connection, may have actually borrowed some of the mechanisms of the pain system to maintain social connections."

Losing important relationships results in far more than loneliness. For example, the long-term health impacts of divorce are well known. The risk of early death for those who divorce is 23 percent higher, according to data from millions of people in eleven countries. This is an increase in risk as large as other serious health risks like heavy drinking, smoking, and obesity.[18] And sometimes you don't have to wait years for the health risks to show up. The day after someone close to you dies, the risk of you having a heart attack leaps to twenty-one times higher than normal, afterward slowly decreasing over time. A broken heart can really break your heart.

GROOMING ON FACEBOOK

Evolutionary psychologist Robin Dunbar has studied the basic elements of how humans live in groups and, as a result, as found "Dunbar's Number." The number is 150; it's the number of people who spontaneously organize in social groups in all kinds of human endeavors, from villages in Africa to departments in Western corporations.[19] You're probably a part of groups that are far larger than 150, and you might have 1,000 friends on Facebook, but you probably have about 150–200 at most whom you really know and care about, and by whom the feeling is reciprocated. Those are your real friends.

The wild popularity of Facebook and other social networks suggests that they are meeting a basic need, but we're not getting any more social because we're wired to the Internet. The Internet gives us a new way to connect, but it does not alter our inner nature. We can afford only so many friends.

Different groups are held together by different types of social interactions, and primate social groups are often held together, in part, by grooming each other. Individuals spend hours with other members of the group, removing parasites, scratching and cleaning each other, and so on. When monkeys groom each other, they build relationships, reduce stress, and establish hierarchies. Grooming helps to build and maintain monkey social networks.[20] Picking parasites from each other is like Facebook for monkeys.

All of this grooming has an upside: investing political capital in the other monkeys and knowing how the social structure works. It does have a cost though. The downside of grooming is that it takes time and energy. Some of the energy is mental in nature, keeping track of the whole range of complex ways all the individuals in the group interact with each other to make sure you're grooming the right friend and doing it the right way, so that you're not upsetting the social order. Dunbar suggests that the size and complexity of our social networks are related to the development of our cortex, which is how he arrived at his famous number.[21] This time and energy spent on grooming might be a solid investment in future relationships, but with only so much time on your monkey hands, and with only so much mental firepower in your skull to keep track of it all, you can groom only so many other monkey friends and family, limiting the size of your social group.

In human groups, the same role seems to be played in part by gossiping. Gossiping gets a bad name, and nasty gossip can cause real and lasting harm. But talking about each other can also be the glue that holds us all together, providing useful information for the group. In parts of Africa with a high rate of HIV infection, gossip about who's infected might have a negative impact on some, but it may help others survive.[22] From ancient times, the information in gossip played an important social role. As populations grew and societies became more complex, humans could no longer know everyone in their environments that well; but we still manage to gossip about friends, as well as celebrities who feel like our friends.

Social networking is the new age of gossip, often aided with our own willing participation. It is the new human form of monkey grooming, the new water cooler for gossip. Rather than picking parasites out of each other's fur, we post about each other on Facebook. But like grooming, logging into Facebook and slogging through all the newsfeeds, posts, pictures, and so on takes time and attention, so we can really talk with only so many of those Facebook friends—150 or so. I hate to break it to you, but the rest of those friends might not really be your friends.

THE SOCIAL MOLECULE

As a consequence of being social creatures, we spend a lot of effort on relationships of all types, from spouses to children, and from friends to coworkers. At the core of these relationships is empathy and trust. Failure to develop empathy and trust can have profound consequences for individuals and society. And at the core of these feelings is the hormone oxytocin.

Oxytocin is associated with a variety of relationships, including finding a mate, caring for a child, and building a business relationship. Oxytocin is also released during sex, perhaps strengthening the bond between two people. While it may seem surprising that a single molecule could do all of these things, oxytocin is not your average molecule.

Dr. Paul Zak at the Claremont Graduate University in California studies oxytocin and its impact on how we trust and relate to each other. Talking about oxytocin in his TEDTalk,[23] Dr. Zak stands on the stage and squirts a syringe of the hormone up into the air, a fine mist dispersing in the bright lights. The hormone mist disappears from sight quickly, but the hormone itself has a lasting and powerful impact on our relationships.

Dr. Zak started studying oxytocin while researching the foundation of morality, looking for answers in our biology. While morality is a difficult trait to get an experimental handle on, Dr. Zak found that he could isolate just one virtue related to morality—trustworthiness—and study it. Trustworthiness is important for many reasons. Countries with a high level of trust and low levels of corruption also tend to be the most prosperous countries. Every deal in business requires trust, and where trust is lacking, it's hard to do business.

"What we do in my lab is we tempt people with a virtue and a vice using money," said Dr. Zak.[24] To study trust, he set up experiments to study how people trust each other with money, using a common scenario often used by psychology labs. Volunteers in a study are matched in pairs of people who can't see or meet the other person. One member of the pair is given some money and told that if he or she gives some of the money to the other volunteer, then the other person will have this money tripled when he or she receives it. Then, this other person can give some of the money back if he or she wants. When researchers do these experiments, they describe the first transfer of money as a measure of trust, while the second transfer is all about trustworthiness, showing that you are someone a person can trust by returning some of the money that is given to you.

A selfish person who doesn't trust his or her partner would just keep all the money and walk away from the game. But in studies like this, about 90 percent of people will send money to the other person. And 95 percent of people who receive money will send some of their gains back again to the first person.

Economists were confused by these results at first. Why do people give money away, sending it back after it is given to them? They're not required to, and they wouldn't suffer any consequences if they didn't; they will never see their partners in the game or interact with them again, ever. The explanation is that people gave the money back because it was the right thing to do. It felt right. "Why not keep it all? That's not what we found," said Dr. Zak. "By measuring oxytocin, we found that the more money the second person received, the more their brain produced oxytocin and with more oxytocin on board they returned more of the money. We have a biology of trustworthiness."[25]

Is it really oxytocin that produces this behavior? The brain is a murky swamp of chemicals in a tangle of neurons, making it hard to single out a particular molecule that is responsible for a complex feeling like trust. So Dr. Zak wanted to go a step further and give oxytocin to people to see what it does to them. Since it's hard for oxytocin to get into the brain, Zak and his colleagues had to experiment a bit. "I used everything short of a drill to get oxytocin into my own brain," said Dr. Zak. They eventually found that a nasal inhaler was better than a drill. With the nasal inhaler, they could deliver oxytocin in mist form into the noses of men and compare their responses in the money-giveaway game. The number of people who gave all of their money to a stranger doubled.

Giving away money and receiving money are not the only factors that release oxytocin. Zak and those working with him found that massage, dancing, and praying also elevate oxytocin. And it plays a role in more than business relationships or sex. When volunteers in a study are given oxytocin, they also increase their donations to charities by 50 percent. There's a video Dr. Zak shows people of a father with his four-year-old son, who has terminal brain cancer. Having measured the oxytocin in their blood before and after seeing the video, Zak noticed changes. When we hear a story like the father-son story, our brain is flooded with oxytocin and we feel for them. (I'm guessing oxytocin is probably responsible for our reaction to stories of a single suffering person and, in turn, for our lack of response to millions of sufferers, discussed in ch. 7.) This feeling for others, this empathy, is the glue that holds us all together.

"Change in oxytocin predicted feelings of empathy, and empathy makes us connect to other people. It's empathy that makes us help other people. It's empathy that makes us moral," said Dr. Zak. "We are social creatures, so we share the emotions of others. If I do something that makes you feel happy, then I share that happiness."

Oxytocin can also make us vulnerable. When Dr. Zak was working at a gas station in the 1980s while he was in high school, a man came in and said he found a jewelry box with pearls, and asked what they should do with it. Just then the phone rang and the man calling said that he had lost jewelry at the station. Dr. Zak (not a doctor at the time) volunteered that they had found the box, and the man said he would be there in thirty minutes and offer a $200 reward to the man who had found it, then hung up. Dr. Zak told the man who found it, but the man couldn't wait, he said—he had a job interview in fifteen minutes and had to go. But if Dr. Zak gave him $100, he could keep the $200 when the other man came to pick up the box.[26]

Dr. Zak gave him the $100, but the other man never came. Dr. Zak had been conned with a classic trick called the "pigeon drop." First the con artist builds trust, and then you open your purse. These tricksters manipulate your feelings of empathy with oxytocin, and it turns out that some people are skilled at this. After testing thousands of people, Dr. Zak's team found that about 5 percent of people don't respond with elevated oxytocin to the same stimuli that most people do. "If there's money on the table, they keep it all. There's a technical word for these people in my lab. We call them 'bastards.' These are not people you want to have a beer with. They have many of the attributes of psychopaths."[27]

Looking at the role of oxytocin in a different type of social setting, Zak studied weddings. Weddings are an intensely social event—we don't usually go to such lengths for just any party. The wedding ritual is about more than having a good time, though; it serves a social purpose. "The wedding ritual connects us to this couple," Dr. Zak said. "Weddings cause the release of oxytocin. We design the ritual to make sure they succeed." Going to the United Kingdom, Dr. Zak attended a wedding and sampled blood from those attending. The person with the highest oxytocin? The bride. The second highest? The mother of the bride. And then the groom's father, the groom, the family and friends, each with oxytocin elevated to the extent of how close their social connections were.[28]

One reason social networking is so wildly popular must be that it makes people feel warm and cozy, giving them a burst of oxytocin like they are hooking up with people they know and trust. One trait of social-networking fiends is that they tend to be a very trusting bunch. I'm on Facebook myself, but I'm pretty guarded about the kinds of stuff I'll say and I log on only rarely. Though some of the folks I've seen on Facebook seem to let their hair down in a dramatic way, exposing way more of their lives than I would ever imagine doing, posting private relationship arguments in the most public of forums for all to see. Similarly, the capacity of Facebook to store and disseminate embarrassing photos is legendary.

When Dr. Zak talks to folks in the media business, time and again they ask him to do a quick experiment to measure oxytocin levels in their blood, often testing the impact of social networking on their oxytocin. The results from testing those at radio stations and magazines may not be a strictly scientific sampling, but the results are consistently positive among these impromptu studies. "Using social media produces solid double-digit increases." One young man he tested for a Korean broadcasting system had a 150 percent increase in oxytocin—a huge jump. What caused it? Interacting on his girlfriend's Facebook page.

There might be a downside to the surge in trust related to social networking. Health experts in the United Kingdom have linked a fourfold increase in syphilis to the growing number of people on Facebook who are using it to connect for casual sex. It seems oxytocin might have lulled them into trusting their Facebook friends more than they should have. While their Facebook friends were open about many things, it turns out that they neglected to post about their syphilis.[29]

FAKE FRIENDS

Television is supposed to be on its last legs, killed by the Internet, but hours of television viewing have increased even with Internet use increasing as well. We want to do both at the same time. There's something about television that still attracts us. That thing is our impulse to make friends, even if the friends aren't real.

When we watch a television show about fictional characters, our intellect knows that these people we watch are fictional, that they don't really exist. At least I think we know that. But deep down inside it feels like something more. It feels like they are our friends, like we really know these people and care about what happens to them somehow. We are absorbed in the intimate lives of the friends of *Friends*, or of Rachel and Finn on *Glee*, because we feel like they are our friends even if the analytical part of our mind knows this makes no sense. They fulfill our innate need to belong, even as the world around us sometimes makes it harder to fulfill this sense of belonging. The television always welcomes us with old friends who never judge us or reject us, even though they might fade away into reruns after a few seasons. When a favorite series ends, fans feel a sense of loss as if they are in mourning. While television may contribute to some social ills, it might also have an upside as well—it makes us feel less alone.

When fans encounter a star, after spending so much time watching her story and getting to know her character, they often want and expect her to be the person they've seen on the screen, the character she portrays. Alina Adams has written extensively about daytime dramas, so-called soap operas, on television, programs that have run for generations in some cases, drawing fans to relate so strongly to characters that they feel like they are a part of their lives.[30] Fans follow these programs and their stars closely and have mourned the recent passing of some of the biggies like *All My Children*. But the impact of these programs in the lives of millions for so many years is undeniable.

"Because soaps are daily, the viewer not only experiences a character's growth, but they are forced to grow along with them as they process the same crisis in real time," said Adams. Seeing these characters on television as part of daily life makes them feel like real people to viewers. "As a result, it's a very intimate, cathartic experience that fills a real need and, often, a void. This kind of perceived closeness makes it almost all too easy to believe that the face on the screen knows you as well as you know them. Which is why, in face-to-face encounters, the temptation is to just leap into conversation as if catching up with an old friend, assuming they know who you are [just] like you know who they are—and [fans must deal with] the shock that comes from realizing they don't. The human brain is, after all, wired to recognize familiar people."[31]

Actress Eden Riegel, who played Bianca, the lesbian daughter of Erica Kane (Susan Lucci) on *All My Children*, talks about how fans related to her character. The program was not just about entertainment or wiling away the hours of the day, it became a part of their lives: "Over time, they began to embrace Bianca along with Erica. And soon Bianca was one of the most popular characters on *All My Children*, even among housewives in Middle America who [had] never even met a gay person before. This story is proof positive that soaps have great power not only to tell dramatic, engaging stories, but also to help change hearts and minds."[32]

THE THING WE FEAR MORE THAN DEATH

Being social feels so important to us that the risk of not fitting in weighs heavily on many and can even lead to social anxiety disorder in some. Many of these people appear shy, with their actions inhibited, and the risk of being socially exposed strenuously avoided. They can go to great lengths to avoid social situations, feeling discomfort while being introduced, having a conversation, or having any attention at all brought their way. However, not all sufferers of social anxiety are the same.

How comfortable people are in social situations can affect the risks they take, but not always in the way you'd expect. Psychology researchers Todd Kashdan and Jon Elhai looked at students and measured their degrees of social anxiety. Kashdan and Elhai then examined the likeliness that these students would engage in risky behaviors such as drinking, taking drugs, having risky sex, or developing aggressive behaviors.[33] You might expect socially anxious people to be risk averse overall. If they're already anxious, why would they go out seeking more risk? But in this study, the socially anxious individuals were looked at further, asking them if they expected positive or negative outcomes as a result of risky behavior. If they think something good might result from taking a risk, the socially anxious group will engage in risky sex or aggression even more so than those who are less socially anxious. The change was not seen for drinking, suggesting that the risky behavior was social in nature—the socially awkward would take the risk if they saw a social benefit was likely, helping them to overcome their awkwardness. The reward for them was social.

Unfortunately, in treatment, these people are often not diagnosed or understood to be different. Being prone to impulsiveness and aggression, they often choose a less-than-optimal solution to social situations. "Our mind is hardwired to respond quickly and automatically to cues of acceptance and rejection," said Kashdan.[34]

While those with social anxiety may be more acutely affected by social situations, they are not the only ones who worry about what people will think of them. Even professional athletes deal with this, some of them having a harder time with it than others. Gregg Steinberg is a sports psychologist working with athletes from a broad range of sports in order to enhance their performance. "I help people master their emotions. The ones who are successful are the ones [who] use their emotions to their benefit," Steinberg said in an interview.[35] Feeling pressure when it's time to perform is not unusual. It's what we do with that pressure that matters. For some the pressure becomes an unbearable burden that holds them back while others learn how to use the pressure to drive themselves forward. "A lot of people feel pressure," Steinberg went on. "The successful ones feel nervous or pressure, but the successful ones harness it into a positive form as a challenge, [to] boost energy levels and focus."

What are athletes thinking about when they're on the field or on the court and their stomach is tied in knots? "Fear of looking foolish, that's the main deal," said Steinberg. "It's the risk of looking foolish. The human condition is to be fearful. If you're a human being and what you're doing is important to you, it's going to make you nervous."

For a vast number of people, standing in front of a group to speak is the worst, most nerve-racking thing they can imagine. The fear can be paralyzing, leading many to avoid doing or saying anything that could draw attention. Maybe this is another odd holdover from those thousands of generations when belonging to the social group was a life-or-death proposition, with people fearing that standing up and speaking may lead to rejection. Today, public speaking is consistently ranked as the greatest fear most people have—ranking higher than the fear of death itself. As Jerry Seinfeld once said, "This means to the average person, if you go to a funeral, you're better off in the casket than doing the eulogy."

Marjorie Asturias of Dallas, Texas, routinely gives talks as the president and CEO of her budding firm, Blue Volcano Media. Today she even enjoys giving the talks, but as a child she panicked at the thought of performing.[36] The fear of public speaking manifests itself in many ways, with some freezing while others are nauseous. For Asturias, the fear made her break into hives. "Growing up, I was so terrified of public speaking and public performance that my legs once broke out into massive hives in the days leading up to my first piano recital at the age of ten," said Asturias. "My aunt was so concerned that she took me to the doctor, who examined me extensively and concluded that there was absolutely nothing physically wrong with me, that the hives appeared to be psychosomatic in nature. I ended up wearing knee-high socks to the recital just to cover up the angry rash wrapped around my legs. Sure enough, the day after the recital they disappeared as mysteriously as they came. I even elected to fail a Reading class in the sixth grade just to avoid giving a five-minute oral book report. And when I had to perform in a school play, I suffered such terrible stomach pain right before I was to go onstage that the teacher had to replace me with another student. Lo and behold, the moment I sat in the audience, the pain was gone."

What are we afraid of? Nobody is killed or maimed because they give a bad speech, are they? Not at any of the talks I've been to anyway. Somewhere deep in our minds the fear must be that we will stand up to speak and fail in front of everyone, and then the whole social group will turn against us. It's as if a bad wedding toast would be met by a hail of bullets or being cast out into the jungle alone. Maybe there was a time when it did.

"In Japan, one is commonly admonished that 'the nail that sticks up must be hammered down,'" said Asturias. "So maybe part of the reason is simply that evolution has taught us self-effacement as a survival instinct, even though the vast majority of us live about as far from the African savannah as we can get."

The fear of public speaking is common, but some people feel it more strongly than others. Dr. Signe A. Dayhoff is a recovered social phobic and a social psychologist. As someone who has suffered through intense fear of public speaking himself, he understands better than anyone what it's like.[37]

"For nearly twenty-two years of terror, I faked it," said Dayhoff. "I was scared to death to speak to more than a couple of people. My mind always went blank. My tongue stuck of the roof of my dry mouth and I couldn't swallow. I blushed, sweated, and trembled. But I was a social psychologist who taught large university classes and had to present my research at conferences and seminars. I did it because I had to."

It got worse though, to the point where he could not even answer the phone. Knowing he could not go on this way, Dayhoff found help. "As I recovered twelve years ago, using cognitive-behavior therapy, patience, persistence, and practice, I found that nearly twenty million individuals at any one time suffer from some form of social anxiety. They fear being negatively evaluated in anything they do; fear being rejected; fear being abandoned."

Having a fear of getting up in front of others to talk or perform doesn't mean that people don't manage. Many people just get out there and do it, nausea, sweat, and all. "Many performers, from Laurence Olivier to Donny Osmond, have this problem but manage to push through as I initially did instead of addressing the problem head on. So that they suffer through every performance or speaking engagement, perhaps with their head over the toilet or breathing into a paper bag before going on stage."[38]

AUTHORITY FIGURES

Authority figures are part of our social structure, people we are supposed to trust. Some of the most common authority figures we deal with are doctors. With their years of technical training and with life-or-death decisions in their hands, we feel the need to trust doctors and hope they can put us at ease. For some of us, a white jacket and stethoscope are enough to win our trust, and in our allotted fifteen minutes of care, we are not encouraged to question them. Unfortunately, though, even doctors are human.

The National Institutes of Health (NIH) estimates that medical errors kill forty-four thousand to eighty-eight thousand people in the United States each year—that's more than car accidents or breast cancer. A 2010 study from the University of California–San Diego found that deaths from medical errors peaked in July of each year, the month when new and inexperienced medical residents move into hospitals and start seeing patients.[39] Furthermore, the studies from the Harding Center for Risk Literacy described in chapter 7 highlight the all-too-human weakness of doctors in dealing with the statistics of mammography results.

Doctors might make mistakes, but authority figures can also use their influence for darker purposes. In the 1960s, a series of experiments conducted by Yale psychologist Stanley Milgram tested the extent to which people would follow instructions from an authority figure, even if this led them to do things that seem obviously wrong.[40] In each experiment, three people were involved. The test subject would be the "teacher," an actor would play the "learner," and another actor would play the "experimenter" who controlled the study. The teacher and learner communicated with sound but could not see each other. The learner would be given simple tasks with words and if they answered incorrectly, the teacher was instructed by the experimenter to deliver electric shocks to the learner. The one in charge, the authority figure, was the experimenter, with the teacher following his instructions. The shocks supposedly delivered to the learner would start small but escalate higher and higher if wrong answers were given. Since the learner was an actor, he was not actually receiving electric shocks but rather pretending like he was.

During the course of the experiment, the learner on the other side of the wall would make increasingly distressed sounds of pain until the actor was screaming and banging on the wall, and eventually falling silent. You would think this would be enough to get almost anyone in the role of the teacher to drop out, refusing to deliver shocks. But in Milgram's experiments, over 60 percent of the participants in the role of teacher implemented the highest possible punishment on the learner, 450 volts, under the prodding of the experimenter, despite the pain and protests of the learner.[41]

People have debated what all this means, but the obvious interpretation is that people have a strong urge to be obedient to authority figures, even when this leads them to do things that seem obviously wrong. Carried to the extreme, this could play a big role in the seemingly evil things people do in wartime, under the instruction and control of authorities.

The Stanford prison experiment is another famous psychological study that examined the power of authority and how it can be abused. Psychologist Philip Zombardo did the study in 1971, recruiting college students to play either guards or prisoners in a mock prison that had been set up in the basement of Stanford University's psychology building. The experiment worked all too well, with the guards and prisoners going far beyond mere role-playing. The guards began to abuse their power over the prisoners, engaging in psychological abuse—and the prisoners allowed it to happen. Although it was planned to run for two weeks, the experiment had to be halted after just six days because it was going too far, getting out of control. The guards were disappointed at the early end to the study because they were relishing the power of their role so much. Years later, when pictures emerged in 2004 from the Abu Ghraib prison in Iraq, the parallel was drawn with the Stanford prison experiment.[42] It seems power really does corrupt, but it needs our help to do it. Corruption is a team effort.

ONE OF US

In the news, we hear a great deal about violent crimes, playing into our fears of human predators in our world today instead of animal predators. While our minds exaggerate gory risks, we have a gaping blind spot for risks like white-collar crime that don't evoke the same primal fear. Although white-collar criminals cost us hundreds of billions of dollars each year, people don't feel threatened at a gut level because there is no physical threat, which in turn leads to a much lower rate of prosecution and conviction overall. Nobody holds a gun or a knife to us, so we don't see it as a threat or, even when we do, we let it slide. According to the National White Collar Crime Center, less than 8 percent of white-collar crimes are reported to the right authorities.[43] For the most part we don't even bother to try.

White-collar criminals do get caught sometimes, of course, including high-profile people like Bernie Madoff. Financial downturns have a way of uncovering Ponzi schemes like his, the financial tide pulling back and exposing them. The surprising thing about Madoff isn't that he got caught but that he was successful at his efforts on such a large scale for so long. He slipped past repeated Securities and Exchange Commission (SEC) investigations and warnings from fraud investigator Harry Markopolos. Many financial insiders stayed away, seeing that the returns Madoff reported were not possible. But right up until the end, most people trusted him. That was the problem—they felt like he was one of them. And that's the trick for white-collar criminals; they win our trust and make us believe that they are one of us, even as they are robbing us blind.

One way Madoff did this was by going to social clubs and being introduced to people. He was powerful, a well-known figure, and he looked and felt like someone they could trust. And the people he met wanted to believe it was true. White-collar crime is a social crime.

"Government has coddled, accepted, and ignored white-collar crime for too long," said Markopolos.[44] "It is time the nation woke up and realized that it's not the armed robbers or drug dealers who cause the most economic harm, it's the white-collar criminals living in the most expensive homes who have the most impressive résumés who harm us the most."

As social animals, we evolved to use subtle cues that tell us who is in our group and who therein can be trusted. For early humans, being a part of the group and knowing who else was in the group was a matter of survival. We look at the hair, speech, clothes, mannerisms, and common cultural touchstones of others as shortcuts to judge the risk others might hold for us.

White-collar criminals exploit these cues to convince us they are part of our social group and within our circle of trust. They often really *are* one from our group, to start out at least. The white-collar criminal could be an ordinary employee who made a wrong move when nobody was looking and soon finds she can't stop. In other cases, the criminal takes pains to avoid sticking out, intentionally blending in and avoiding drawing attention his way.

Jim Angleton worked as the CFO for the United Teachers of Dade in Miami, Florida, for only a short time before he caught the CEO Pasquale "Pat" Tornillo embezzling more than $5 million from teachers' union dues.[45] Tornillo was living far beyond what his salary alone could have provided, expensing lavish trips and more on the corporate credit card. Soon after he started, Angleton had compiled six volumes full of checks and receipts, evidence that he used to help bring Tornillo to justice.

The evidence against Tornillo was not hard to find, but for a long time nobody was looking. Even with the evidence in hand, taking down an established figure like Tornillo proved extremely difficult. He had done his work well, building trust in the community for so many years that he was a well-respected name in Miami. Confronting him as a whistleblower was not easy for Angleton. He faced death threats, spent $150,000 of his savings on legal fees, and went looking for work for an entire year after the whole ordeal was over. Today he is the president of Aegis, a business-intelligence company.

When talking about these boardroom criminals, Angleton suggests some of the traits they use to win our trust. He starts with their own personal appearance, of which they take great care, with their haircut, well-groomed nails, expensive suits, ties, watches, colognes, shoes, glasses—all of it designer items and a display of "conservative" wealth. They live the "country-club lifestyle," sponsoring the right charities and loving to get their picture in the paper about their philanthropic efforts. They drop the right names and walk in the right circles. They "lull you into a secure sense you are in the company of a trusted person," said Angleton in an interview. But seldom do they talk themselves up. That might raise our suspicion. Instead they get others to talk them up, doing the work for them.

Tornillo met Angleton's description to a T. As did Madoff and many others. These activities described on Angleton's list might not be all that different from successful businesspeople, however: dressing for success, networking, and building trust. It's what they do with the trust that is the problem.

Being social is a powerful influence that runs throughout our lives. It's part of many problems we face and can be a strong influence to sway, nudge, and shove us in the right direction as well. Like many tools, how we use our social nature, or how it uses us, is ultimately up to us.

CHAPTER 9

WHAT DO CAVE DIVING, NAKED SKYDIVING, AND ENTREPRENEURSHIP HAVE IN COMMON?

The Delicate Dance of Risk and Reward

DEEP, DARK, AND WET

J ill Heinerth loves exploring caves, but not just any caves.[1] For some explorers, traveling deep in the earth through winding caves is the great adventure, but Heinerth wants more. She loves exploring underwater caves, deep in the farthest reaches of watery caverns where few, if any, have traveled before. Even as humans have explored the highest mountains and outer space, these submerged caverns remain largely a mystery, a different world deep within our planet. This world is so difficult to get to that it has taken the life of many who have dared to enter, making cave diving one of the riskiest things you can do.

But that doesn't keep Heinerth away. While most of us go to great lengths to avoid such great risks, people like Jill seem drawn to it. Like many explorers, she has lived with the risk of death on almost a daily basis, slipping close to the "other side" many times. And while she knows that being smart about risk is crucial to survive in her line of work, explorers like her seem to thrive on pushing risks as far as they can. Explorers are part of a unique class of people who seem not only to tolerate risk but also to thrive on it. For them, a life without risk is not really a life.

Thinking back to how she started down the path to becoming an explorer, Heinerth remembers a key moment in her life, a turning point: "As a young university student, I fought off a burglar in my house with a small X-acto blade and a paring knife from my drafting table," said Heinerth. When she fought off that intruder, she proved to herself that she could face death and come out intact. If you do that, you can do anything. "That incident empowered me. When I heard the burglar in my house, my first instinct was to pull the covers over my head and hide. By the time he entered my room and tried to attack me, I had somehow summoned the courage to fight, win, and live."[2]

With time, the seed grew, leading her into the deep, dark world of underwater caves. "I started diving after university," said Heinerth. "It was something I always wanted to do. I grew up weaned on Jacques Cousteau episodes and the Apollo launches. For me, cave diving is the earthly equivalent to exploring space . . . a last frontier."[3]

Few of us are even aware of these underwater caverns that lie under the surface. They can extend for miles, deep underwater in darkness, with confusing, tight paths intertwined in a maze leading back to the surface. There are rivers, lakes, and cavernous passages traveling so far into the earth that nobody really knows their true extent. When a diver is far inside the reaches of a cave, breathable air can be so far away that any problems with breathing equipment can be fatal. There's little or no margin for error.

Jill has many years of experience with the equipment and the caves, and she does everything she can to manage the risk, short of staying away. "The ominous doorways of underwater caves repel most people, but I am attracted to the constricted corridors, squeezing through, relying on delicate technology for every sustaining breath," she said. "A bad decision at work could cost me my life. My name and legacy would be added to the long list of cave divers who have perished in the seductive blackness."

Knowing the risks, Heinerth controls her fear but also uses it to keep herself sharp and alert. "Per attempt, some have said that cave diving is the world's most dangerous sport. Fear keeps us in check. We previsualize our dive before entering. We try to imagine the things that might go wrong and the solutions to them. Still, we have to keep a certain amount of fear in our hearts in order to come home at the end of the day. The fearless usually die in this sport. The important thing is that you do not allow the fear to make a hostile takeover of your brain. You have to keep the emotions down so that there is still room for reason to lead the way out. In a crisis, you need to make the next good decision to get a few feet closer to home."

Heinerth has dived in caves around the world, but one of the most unusual was diving in an iceberg in Antarctica, the B15 iceberg off of the Ross Ice Shelf. As Heinerth and her diving colleagues worked their way forward in the subfreezing water, things took an unexpected turn. "We were underneath the berg when the current turned and it was so strong we were having a tough time swimming out. Every fin kick forward was one slip backward. Using perhaps the highest workload I have ever put out, we eventually made it back to the entrance crack and to our decompression stops [to keep bubbles from forming in their blood, divers must stop along the way while coming to the surface], but they were greatly extended by our delay. Some three hours later we were finally able to emerge from the 28-degree water, half-frozen after a dive that had tripled in planned length. Several hours after the dive, the entire berg rolled and shattered like an ice cube dropped into a drink. It exploded and turned into a square mile of brash ice floating on the sea."

Heinerth and those who climbed out of the water with her were stunned and exhausted. "At the moment of trying to get out of the iceberg I was thinking 'Wow, this is how it goes.' Your life does not flash before your eyes. On the other hand, all you can think about is working harder than you have ever worked to live."

"As I climbed the ladder to the boat, I said, 'The cave tried to keep us today.' And that was it."

That was it. They had lived to dive another day . . . and another, and another. And love every day of it, way out on the edge. "Survival is rather empowering and it leaves one to treasure every moment and concentrate on living fully. I always told my Mom . . . I am not afraid of dying, but I am very afraid of not living . . . fully. We have such a short time on this planet to do something that matters."[4]

Life. And death. Maybe one doesn't mean much without the other. Sometimes it's only by taking risks that tempt death, pushing right out to the edge, that we feel like we're fully alive. By taking a risk and tempting death we can find out what really matters to us in life.

LIKE MOTHS TO THE FLAME

Being an explorer has its rewards, but there are easier ways to earn a living. Explorers take these risks because they want to. Going out of our way to take risks such as these might seem like odd behavior if we are risk-calculating machines, but we are not. Our risk taking stems from the wetware in our skulls, a few pounds of infinitely complicated neuronal mush. And the result of this meshwork deep in our brain can be surprising, such as the connection between taking risks and feeling good.

Our brain responds to a variety of stimuli that make us feel good, often for obvious reasons. Eating feels good and can produce a very rewarding and satisfying feeling, and for good reason: eating helped countless generations of animals survive over millions and even billions of years, so there's a strong evolutionary pressure to reward animals (including humans) that are successful in their quests for food. Sex and love make us feel very good indeed, for a different reason than food but still with a strong biological rationale. Evolution has reinforced the rewards for sex and love since these help us to reproduce, passing on our genes by associating reproduction with very rewarding feelings.

If a patient in an MRI scanner is experiencing these inputs, his brain will respond and the pleasure centers will light up on a functional scan. But sex, love, and Twinkies are not the only way to turn on that rewarding feeling, and here's where things take a turn for the weird.

For many people, taking risks makes them feel good. In turn, their brain's reward center lights up, using some of the same neural pathways as the patient's brain when he experienced the rewards from food, sex, and love. Some risk takers respond this way because of the way risks and rewards are connected in our brain. There are a wide range of things that feel good, and different things feel good to different people. But one thing that food, sex, and gambling have in common for those who enjoy them is that they cause dopamine to be released in the brain in the reward center. In parts of the brain such as the nucleus accumbens and the frontal lobes, doing something that feels good causes dopamine to be released. And we really like it when this happens—we like it a lot. Different people get their reward center triggered by different things. Different strokes for different folks. You might not be into gambling or cave diving, but it seems like one way or another we all need that reward center to light up. So much so that we'll do almost anything to get this feeling, even if it means going to risky extremes.

The brain is a complicated place, and nothing really acts on its own in there. Dopamine release is controlled by a variety of nerves and neurotransmitters, including cannabinoids, serotonin, and others. But all roads seem to lead back to dopamine and its role as the feel-good neurotransmitter of the last billion years.

Controlling how dopamine is released in the reward center seems to be at the center of drug abuse and alcoholism. You might think that people who get an unusually strong reward from risky behavior or drugs are the one who would indulge in these the most, but this seems to not be the case. In fact, those who take risks seem to have the opposite problem. While we all release dopamine to get that natural high, some of us release more than others. But rather than getting too much dopamine, risk takers seem to have a lower-than-normal response in their reward center. They engage in gambling, extreme sports, or taking drugs in order to boost their reward signal to the same level that others achieve with a lower level of stimulation in more mundane living. They are going to extremes just to feel normal.

There's even a name for what these risk takers are experiencing: reward deficiency syndrome (RDS).[5] People lacking the normal level of reward feelings in their brain will seek out risks like compulsive gambling, drugs, or risky sex to boost it.

THE STRANGER SIDE OF PARKINSON'S DISEASE

Many of the body's molecules have more than one job, taking on multiple roles depending on when and where they are working. Dopamine is one of these moonlighting molecules, doing more than just making us feel good. Dopamine also lies at the core of Parkinson's disease.

Parkinson's is a progressive neurological disorder in which sufferers gradually lose control of motor neurons and the corresponding muscles controlled by these neurons; thus, eventually, they can no longer move normally. In recent years Michael J. Fox and Muhammad Ali have raised awareness of Parkinson's as a serious disease that millions of people face. As time passes, Parkinson's transforms the lives of those afflicted, leaving them unable to do the normal activities of daily life they once enjoyed. At first, the symptoms are not as intrusive and include a slight shaking of the hands, but with time the symptoms get worse until movement of any kind becomes almost impossible to control without medication.

A variety of drugs are used to boost dopamine signaling in Parkinson's patients, giving them more control over their movements. Levodopa has been used for decades to reduce the symptoms of Parkinson's, mimicking dopamine and restoring greater control over movement for patients (for a time, at least). But that's not all the drugs like levodopa did for Parkinson's patients. Over time, reports started to surface of Parkinson's patients who were taking these drugs and engaging in very unusual activity. Most patients respond well to this class of drugs, regaining more normal motion, but some of them also took to compulsive gambling, risky sex, or other risky behavior they had never engaged in before.

Dr. Leann Dodd at the Mayo Clinic in Rochester, Minnesota, found that Parkinson's patients taking drugs for their illness suddenly got into compulsive gambling after a lifetime of careful living.[6] Patients who never gambled in their lives suddenly had been found to have gambled and lost thousands or tens of thousands of dollars, sometimes losing everything. In the United Kingdom, seventy-two-year-old former police officer Morton Wylie sued for damages after taking an experimental new Parkinson's drug.[7] Wylie had gambled little in his life, buying a lottery ticket on occasion, but once on the Parkinson's medication he gambled away his family's savings and then some.

Hypersexuality is another side effect in some patients.[8] People who lived quietly their whole lives suddenly were found making frequent inappropriate sexual advances, visiting prostitutes, cross-dressing, or developing a taste for pornography. They develop intrusive sexual thoughts and behaviors they find hard to control and might not want to control. They might shop and eat compulsively, unable to restrain impulses that they had previously dealt with their entire lives without much effort. Some patients find they like these side effects, asking for increased dosages of the drugs.

Failure of the dopamine system to respond normally to the world around us may play a role in a tendency toward drug addiction or risky behavior, or it might also lead people to pursue the cutting edge as explorers, leaders, and entrepreneurs. It's a cruel trick that food, sex, money, and drugs all provide the same reward in our brain, but we did not evolve this trait because early hominids enjoyed gambling. Odds are we possess this trait because it drove our ancestors to take risks when they needed to in order to survive.

DREAMS OF FLYING

Skydiving can be a thrill in itself, but for some people, jumping out of an airplane is not enough. For these, there is BASE jumping, where thrill seekers wear a parachute and jump off high structures, such as buildings, cliffs, bridges, or towers. If skydiving is risky, then BASE jumping really kicks it up a notch. We can judge the risk of BASE jumping based on some statistics: one in 2,300 jumps will result in death, which is forty times riskier than hang gliding and fifty-three times riskier than skydiving.[9] Those who go BASE jumping are no doubt aware of the risks involved, but nobody is forced to BASE jump. They do it because they love it, because something inside drives them to go further than the rest of us, doing things that make most of us shudder in fear.

Roberta Mancino is one of the world's leading BASE jumpers, successful both as a model and as an extreme athlete.[10] She started skydiving when she was nineteen, but she had dreamed of flying for years. As a child in Italy she would ask her mother to take her to the park where she would see the formations of birds flying to leave with the changing of the seasons. From there she played with flying toy helicopters and eventually, when she started skydiving, she finally felt like she was realizing her dream. While in the air in free fall, controlling her movement with the position of her body, she feels she is flying like in her dreams. Flying is an ancient human dream, but Mancino takes this dream to another level, flying with only her body, her parachute, and the suit she wears . . . and sometimes she forgoes the suit to take the occasional naked skydive.

BASE jumping has taken Mancino's dream of flying to another level. After logging thousands of skydives, she tried BASE jumping from bridges, towers, and cliffs, logging over seventy BASE jumps so far. From there she has gone to diving with a wing suit, an outfit worn while diving that creates winglike surfaces stretched out from the arms and legs. And if this isn't risky enough, she has started to do proximity diving, in which she dives and soars as close as possible to mountain peaks or other objects. With all of these, she is pushing right out to the cutting edge of risk.

"My favorite jump is with the wing suit," Roberta commented.

"It's best from cliffs. I love buildings too—they're so exciting!" We talked on the phone, though I imagine her face glowing with the excitement of diving and flying through the air this way, as close to flying like a bird as humans are likely to get.

Her greatest notoriety has come from those occasional naked skydives. While using the same safety gear as before, skydiving naked does have its downsides. It's cold jumping out of a plane at that altitude, and most of us would feel even more vulnerable than skydiving with clothes on. It does make quite a sensation, though.

All of this excitement and her love of flying come with a risk, which Roberta knows quite well. We talked just days after Roberta's boyfriend Jeb Corliss, another famous BASE jumper, was hurt doing a proximity dive in South Africa, diving from Table Mountain and crashing into its side, fracturing the bones in his legs in multiple locations. Jeb has logged a long series of spectacular dives himself, using the wing suit to proximity dive, soaring just feet away from rocky ridges in the Italian Alps. Jeb had a previous brush with death during a dive that sent him plunging into a waterfall. The hospital reported after his most recent accident that he's likely to face reconstructive surgery, time in a wheelchair, and legal charges for doing the jump without a permit. While Roberta and I spoke, he was still recovering in the hospital.

Roberta's experience helps her avoid the worst-case scenarios, and if something is too crazy, she won't do it. Her definition of crazy might be different than the way you or I might define it, but knowing what you are doing always helps; the riskier the activity, the more that experience matters. Those who have accidents in BASE jumping often have too little skydiving experience. "All of my students want to do just like my videos right away," said Mancino. But she did thousands of skydives before BASE jumping. What she does in those videos is not something you should do on your first or even your one hundredth time out.

While doing her best to dive safely, taking a risk is part of why she loves and dreams about flying. "Risk makes it exciting, more enjoyable," said Mancino. "You have to move exactly in BASE jumping, with no room for mistakes, and sometimes you're doing something illegal, so there's more adrenaline. When you're down, you have to run away from the police. For events I did in Malaysia, jumping from a building, it was a very dangerous dive and you need to be perfect. It is a big challenge, but I like to feel the challenge, the challenge to yourself. It's about how far you can go, about how much courage you have."[11]

"On my first jump when I landed, I could not feel my legs, shaking because I was so excited, jumping up and down, shouting and cursing," said Mancino. "It feels like you are a superhero, like nothing can hurt you, that nothing can be a problem."

I can picture it, almost feel a hint of it even, tingling. It must be quite a sensation.

DESPERATELY SEEKING SENSATION

The reason that Jill Heinerth and Roberta Mancino stand out in our minds is that they enjoy seeking out much greater risks than most of us would voluntarily subject ourselves to. We all factor in and accept some level of risk when we get out of bed and get in a car in the morning, but we get out of the way when a greater risk comes along rather than running to embrace it. There's something very different about risk seekers like this, not just in what they do but also in who they are. We can see this difference when we look at the brain of risk takers in action.

One trait many risk takers seem to share is that they are sensation seekers. They love strong sensations. They don't test the water—they dive in. If they go to a Thai restaurant, they want spicy-level eleven. They love adrenaline, and they love novelty, and they want a lot of it. And often sensation seekers are prone to drug abuse or other negative risky behaviors.

Thomas Kelly, a psychology researcher at the University of Kentucky, subjected volunteers to personality profiling and MRI brain scans.[12] The profiling questionnaires reported if people were sensation seekers and risk takers (or at least if they exhibited these traits while answering questionnaires). When they were lying in the scanner, they were shown a variety of pictures, some not too exciting and others fairly shocking. In response to violent, erotic, or other images that would provoke a strong response, people who were risk takers (according to the questionnaire) had specific regions in their brain light up differently than those who were risk averse. In response to the more arousing pictures, risk-averse individuals' frontal cortex lit up; this is the area that regulates emotions, keeping a lid on things. A result of this is that risk-averse individuals are better at blocking impulsive but risky actions.

Gambling is not for everyone, but those who really love it can't get enough. In fact, when you go to Las Vegas, there are no windows and no clocks on the gambling floor. The casinos know that people who love gambling will lose all sense of time, submerged deep in the reward signal of taking a risk and putting money on the line.

Other risk takers experience the same thing. Pathological gamblers might start out small in their gambling but will escalate their habit over time until it destroys their finances, jobs, homes, and relationships. It consumes them. But what they are pursuing is about far more than money. Looking at their brain, researchers studied pathological gamblers' responses to a simple card-guessing game while their heads were scanned with an MRI.[13] The task was picking one of two cards to find a red card. If the card was red, they would win one euro; if it was not, they would lose one euro. The stakes were not large and the game could not be simpler, but the result was clear. Twelve pathological gamblers were compared to twelve individuals who were similar in many respects but were not pathological gamblers.

One of the first things researchers noticed was that both groups' reward center in the ventral striatum lit up when they won— everybody loves winning. But the pathological gamblers did not respond as much. It's not that they find gambling more rewarding than other people; in fact, it's the opposite. And since they find it less rewarding, they have to go to greater lengths to achieve the same level of reward as nonpathological gamblers.

BUSINESS THRILLS

The world of business startups is far from a steady job, and entrepreneurs have a reputation for taking risks, putting everything on the line to be the next Zuckerberg, Jobs, or Gates. If you don't make it, you could end up on the bottom of the heap, looking through Craigslist. Even if you do make it, your business could quickly be made obsolete, becoming the next Netscape, MySpace or Kodak. The ride from the bottom to the top can be fast, but the trip back down can be dizzying as well.

Surviving in this environment, and succeeding in it, may require a certain type of person willing to take some risks. And a certain type of decision making.

Among the many decisions we make, some are hot and some are cold. Cold decisions don't get us very excited. They can be analyzed without a lot of emotional baggage getting lugged along and swaying us one way or another. Solving a jigsaw puzzle is a cold process. Risky decisions about the future of your business, on the other hand—with millions of dollars, your reputation, and the jobs of your employees at stake—are hot decisions. Our brain handles hot decisions differently than cold ones. To handle a hot decision, our emotions are involved as we weigh the risks and rewards. Deciding how to pitch your business to investors is a hot decision, loaded with emotion whether you succeed in your pitch and get the funding to get off the ground or you fail to secure investors and continue grinding it out.

Barbara Sahakian, professor of clinical neuropsychology at the University of Cambridge, and colleagues compared hot and cold decision making in two groups of people: managers and entrepreneurs.[14] Cold decision making was measured in a simple puzzle, rearranging balls in holes to put them in a certain order. No money or other stakes were involved, and solving the puzzle did not get anyone's adrenaline going. Hot decision making was measured in a guessing game called the Cambridge Gamble Task, in which people were presented with red or blue boxes and had to guess where something is hidden, gambling on the outcome.

Managers and entrepreneurs performed equivalently on the cold decision-making test, the puzzle. And they both were able to make sense of the Cambridge Gamble Task. However, entrepreneurs bet more on the task, taking greater risks. Supporting this result, it has been recorded that, on average, entrepreneurs score higher in impulsiveness on personality tests. While most people become more risk averse with age, entrepreneurs bet like young people.

The role of impulsiveness in risk taking by entrepreneurs is connected to specific brain regions and perhaps even specific neurotransmitters—and we're back to dopamine again. "Using single-dose psychostimulants to manipulate dopamine levels, we have seen modulation of risky decision-making on this task. Therefore it might be possible to enhance entrepreneurship pharmacologically," wrote the authors of Sahakian's study. An interesting thought—a pill for entrepreneurship. And maybe we can come up with another pill to cure entrepreneurship for those who are habitual business starters.

A professor of neuroscience at Johns Hopkins University School of Medicine, David Linden has proposed that many leaders have a similarity to addicts in their attenuated responses to reward. Linden is the author of *The Compass of Pleasure: How Our Brains Make Fatty Foods, Orgasm, Exercise, Marijuana, Generosity, Vodka, Learning, and Gambling Feel So Good*.[15] Addicts are not merely prone to moral weakness but also suffer from a biological derangement deep in their brain, seeking the same reward we all seek, but forced to go farther to get it. Entrepreneurs may have the brain of a junky, but they get their highs from the risk of starting a business rather than a drug they would smoke, eat, or inject.

Our environment today is far safer than the world of our ancestors, and there's not always the need to take great risks to survive. But our inner urge that drives us to take risks is still there. If the world of our ancestors was a risky world, then being able and willing to do risky things may have been important for some to survive. The humans who set out over time into unknown areas, migrating to populate the globe, were taking a greater risk than Columbus. At least Columbus knew there was an India somewhere out there, even if he was confused about its location.

This ancient pursuit of risk can still serve us well at times today. We might not all be skydivers or cave explorers, but we all have risks we face in life. Sometimes the ancient urge to take risks helps us get up the nerve to ask someone out on a date or to call about a job. Taking a risk might be the thing that makes it all worthwhile.

CHAPTER 10
LIVING IN A RISKY WORLD

RISKALYZE ME

After looking at risk in so many different ways, one thing is clear: a big part of the story is not found in the world around us but rather inside each of us. There are common influences that act on all of us, like our aversion for losses and our need for control, but we each handle these influences in our own way.

Money, for example, is not just pieces of paper or a piece of digital information we use to buy things. Money is the biggest source of marital discord for a reason—because money has strong feelings attached to it, often related to fear and risk. Doing a better job at dealing with these strong feelings might help us handle our money and investments better.

While large corporations have whole teams to look at risk and how to manage it, most of us don't have access to these kinds of resources to manage risk in our own lives. Instead we rely on the little guys upstairs in our brain. Riskalyze, a start-up company based in Northern California near Sacramento, has a better way, helping individuals get a better understanding of their own personal relationship with risk and how this affects their investments.

While we think of risk in our investments as something out

there in the market, there are really two components, the risk of the investments and how we perceive those risks. Riskalyze is working on the second part. "Our technology allows people to quantify their risk tolerance," said Aaron Klein, CEO of Riskalyze.[1] "It helps you quantify your risk tolerance as you see it, and align your investments with your risk tolerance."

The system works by walking individuals through a series of questions about a variety of scenarios in which they can either sell an investment or hold onto it and have a chance of either losing the money or seeing the investment grow. If you invested $1,000 and it is now worth $1,100, would you sell the investment or hold it if there's a 50 percent chance of losing everything and a 50 percent chance of seeing it grow to $3,000?

"It usually takes about twelve questions to dial in and understand when you would choose certainty and when you'd take a risk," said Klein. There's no single right answer to the questions they ask, only the answer that is right for you, reflecting your own individual risk tolerance. Your answers are probably different from what Warren Buffet would say, or what I would say, but whatever your answers are, they help you understand your own individual risk tolerance and use this information to invest in a way that you are comfortable with.

When people go through the Riskalyze system, their answers are commonly risk averse, staying away from all-or-nothing options. That's not unusual. The prospect of losses usually weighs heavily in the minds of almost everyone, having much more of an impact than the prospect of gaining money. In the aftermath of the financial collapse of 2008, I'm sure many people are rethinking how much risk they want in their investments or can comfortably tolerate without agonizing over the daily bumps, jerks, and twitches of the markets.

Many factors influence how people respond to the questions

Riskalyze poses. "When we see that the Dow has cracked 13,000 we get more optimistic, and if the Dow falls 600 points today then probably you're not," said Klein. Even little things at home can flavor how we see our investments. "If your boss or your wife looks at you funny, then maybe you're not feeling good and not making any large purchases."

The evaluation helps investors understand their own risk toleration, giving them a mirror to understand themselves better, but it does not eliminate risk. "We'd all love to be able to see the future, to have a magic button we could push that would tell us which investments will go up or down," Klein explained. "But then everybody would push it and it would not be accurate anymore. The magic button does not exist."

Darn it.

As I've learned about risk and how we deal with it, I've wondered if there might be a better way to do things than relying on our gut instincts. We've seen plenty of ways in which our gut gets things wrong. When it comes to large numbers, rare events, and many other situations, the gut becomes dramatically unreliable, and feeling our way through risks can lead us down a street to a part of town where we'd rather not be.

When I spoke with Klein, I wondered if we could develop a similar tool to riskalyze other parts of our lives, maybe our love lives, for one. As we've seen, love is a risky business, and having a better handle on love is probably at least as valuable as having a tool to understand our risk tolerance in investments. I'm pretty sure there is no magic button for this one either, though.

Darnit, darnit, darnit.

SOMETIMES YOU'VE JUST GOT TO SAY . . .

Jennifer Steil was working in New York as a senior editor for a national news magazine when she got an e-mail from her high-school boyfriend asking if she wanted to come to Yemen to train journalists.[2]

As one of the poorest countries in the Arab world and frequently wracked by conflict, Yemen does not have a great reputation according to most people in the United States. Steil was intrigued by the offer, but while the adventure appealed to her, it didn't feel like something she could really do. It seemed impractical, so she compromised. "I said I couldn't just quit my job and run off to Yemen, but that I could come for three weeks to do a training for the staff of the *Yemen Observer* paper," said Steil. "So I spent my vacation in Sana'a, training an amazing group of journalists, half of whom were women. I have never taught such an eager, enthusiastic, ambitious group of people! They were desperate for knowledge. At the end of my three weeks, the Yemeni owner of the paper asked if I would be willing to come back and take over the paper as editor in chief for a year."

Now, Steil is a bit of a risk taker, not one to shy away from living life on the edge. Perhaps this is a common trait in journalists. "I talk to strangers and I stay in the homes of people I don't know. I've climbed down the Grand Canyon in a back brace two weeks after getting hit by a car, fainted in a Costa Rican brothel, and gone years without health insurance."[3] So it's not that Steil hasn't taken a few risks in her days, but ditching everything to go work in Yemen for so long seemed like sailing-off-the-edge-of-the-world sort of stuff.

Still, when after three weeks she returned to her New York life and her drab Midtown office, she realized how tired she was of the same old routines. For most of her life she'd been longing to live abroad, though Yemen hadn't actually been at the top of her list. But maybe this is what it would take to launch her out of an ordinary life. "I am crazy to turn down this offer!" she thought, reconsidering. "Who else is going to simply hand me the reins of a newspaper?" She went back to Yemen and didn't stay for a year. She stayed in Yemen for four years. "And not until the last year did I get held up at gunpoint," said Steil.

Yemen was not what she expected. "Before I moved there, I felt frightened of Yemen. Everything I had read about it was negative. And the US State Department website basically says 'Don't go to Yemen or You'll Die!' In perhaps not those exact words, but close. And there has been a lot written about kidnappings in Yemen. So I was nervous."

Yemen was not Manhattan, but it was not random anarchy either. Despite the challenges and differences, she quickly adapted. "It was amazing how quickly I lost all of my fear. Everything became normalized for me—the fact that all the women were covered, all the men wore daggers on their belts, that AK-47s were a common accessory, etc. I felt utterly and completely safe walking the streets of Yemen. Yemenis are the friendliest people I have ever met. I would be introduced to someone, and he would say, 'Hi, my name is Mohammed, won't you come to my home for lunch with my family on Friday?' And I would always go! It seemed rude not to. And they just took me in as part of their families. So I was able to be reassuring to my friends and family that I was not in a dangerous place. My parents even came to visit me."

Yemen was not without its dangers. While she was living there, a group of Spanish tourists were killed in a bombing in the town of Marib, not far away from Sana'a. Only two weeks earlier, one of her housemates had visited there. "It kind of drove home that it could have been us. But I was busy organizing newspaper coverage of the attacks, so that kept me from thinking about any personal risk."

Bombs are a scary thing—sky high on the riskometer. They are malicious, fast, out of our control, and can do quite nasty things to the body, such as killing it. But that didn't stop Steil. "Later I traveled across the country to the east, to Hadramout, with my Arabic teacher. We drove through Marib, where the bombing occurred, and just as we entered the city I got a text from the British ambassador, by then my fiancé, letting me know that another car bomb just exploded in the city and we shouldn't stop there. We did stop anyway—we needed lunch! And did a quick tour of the sites. I think I felt that as long as I was with a Yemeni, I was okay."

Eventually, after four years, Steil did leave Yemen, though she did not return to New York. She moved to London with her fiancé, whom she had met in Yemen, and with their child. While in Yemen her life had changed. She had gotten far more out of the experience than she had ever imagined. Not only had she had the adventure of living in Yemen and working with its people, but she had also met her life partner, become a mother, and chronicled her experiences in a book, *The Woman Who Fell from the Sky*. Taking the risk and leaping into the unknown had changed her life.

"I took this enormous risk—I gave up my job in NYC, a good salary, health insurance, a boyfriend, my apartment, etc., to run off to Yemen and run a newspaper," said Steil. "But in return I had the most amazing four years of my life! So Yemen was good to me—I got a book, baby, and life partner out of the deal. I don't think I could really ask for more. So I do think that dramatic risks can have dramatic pay-offs."

So do I.

RISKING EVERYTHING

In this book I've focused on the many ways that we fail in our dealings with risk. The list is long. Those who think they'll never get sick and don't need health insurance have often found themselves suddenly bankrupted by illness. Employees of Enron found out the hard way that putting all of your money into company stock really is a risky investment. There's no lack of failures like this.

But the failures are not the whole story. Not by a long shot. We also get things right a great deal of the time. Our gut instincts, intellect, and experience can make mistakes (and sometimes they're whoppers), but we also do a surprisingly good job a large percentage of the time.

"Actually the affective system is a very rational system," said psychology professor Paul Slovic when I spoke with him.[4] "It usually works. It's the way we navigate through our day. Driving your car, you get affective cues about when it's okay to pass or speed up, with your feelings guided by experience, whether doing these things would be good or bad. It's a highly sophisticated system even if sometimes it breaks down."

When things break down, the consequences can be severe, but if our means of dealing with risks didn't work most of the time, we wouldn't be here at all.

One way to deal with how we see risks is to say that there's not much we can do about it. Why fight human nature? We are who we are, right? That's not really what I'm trying to say at all, though. While there are many influences that nudge, sway, and push us in one direction or another, we have the uniquely human ability to choose our own paths, to avoid going with the gut or biology and find our own way.

Having a better understanding of ourselves and why we respond the way we do can help us make better choices, avoiding the big problems and dealing better with the small ones. We have great tools in our biology, our intuitions, and our analytical abilities. By being aware of the many influences on our decisions about living and risk, we can do a better job overall at navigating our world and working together to manage the world's risks.

There are some common patterns that emerge after looking at risk from so many different angles. One of these is the huge role that anxiety and avoidance of it plays in our lives. It drives us to seek control in a car, drives us to deny heart attacks, and makes us break into a sweat at the thought of giving a speech. Avoiding anxiety is a huge motivator for many of the things we do, even if avoidance isn't always entirely productive. Fear has its place and can be quite useful when being chased by a water buffalo, but constant gnawing anxiety seems to send us running in the wrong direction. All too often, avoiding anxiety becomes the goal rather than fixing the risk that is causing it. Finding better ways to deal with anxiety in our lives and its underlying causes seems like a better way to go.

The importance of being social also shows up over and over again. Consider what leads us to prepare for earthquakes: watching the neighbors fasten their bookshelves to the wall. We embrace the fight against climate change, or deny the problem, by banding together with others who do the same, sharing our beliefs. Our social nature leaves us vulnerable to con men and authority figures but also allows us to work together to take on the great challenges our species faces in the present day. From the day we enter the world until the day we leave it, we take on the world's risks together, whether or not we like it.

While this is by no means a book solving all of the world's problems, there are a few strategies worth noting that might help us all. A few of these encountered along the way in this book include:

- Think of the long-term.
- Link long-term goals to short-term personal pain and payoffs.
- Make solutions social.
- Help people to feel in control with solutions.
- Put a name and a face on problems (in other words, don't be a statistic).
- Deal with the causes of anxiety rather than anxiety itself.
- Talk about values rather than arguing facts.
- Tell a good story.
- Try incentives rather than just punishments.
- Don't let those old biological influences get the better of you.
- Don't panic.

People devote lifetimes to and write volumes on managing risks, and your options are by no means limited to those strategies touched on here. If you're dealing with a risk in your life and your work, and you're working with others who face risks, these simple ideas might be worth a thought.

At the end of the day, there's more to life than avoiding risks. Taking risks is part of life. Leaving the security of a job to launch out into the unknown and start your own business is a risk. Having kids is a risk; falling in love is a risk. Sometimes these things hurt. Businesses can fail. Kids grow up and move away. Marriages and relationships can fall apart. Bad things happen to almost everyone at some point along the way, but that doesn't mean we're going to stop living. To avoid risk entirely for fear of failure would mean missing out on all the best things that life has to offer, which is an even greater risk.

So there we are. Krishna, the shark-attack survivor working to save sharks; Roberta, the naked skydiver who loves to fly; Jennifer, the journalist dodging bombs and finding a new life in Yemen; and, last but not least, you and your life. Are you taking too many risks? Or not taking enough? Nobody knows the answer but you.

Imagine yourself strolling on the African plains a million years ago, *Homo erectus*–style. It's a warm spring morning as you stride through the tall grass, some antelope grazing nearby. The hyenas are at a distance, far off, so you relax a little. You're looking for a place to get a bite to eat, as always, but you're not desperately hungry, and you can see others in your family nearby doing the same. You wave, and they wave and smile back. Life is good. Maybe you'll go down to the watering hole later and throw rocks at the hyenas—that'll show those bastards.

Now look around where you stand (or sit) today. Wherever you are, the risks you face are vastly different from a million years ago. We have cars, mortgages, and jobs. We have climate change, obesity, and mad-cow disease. But the fundamental problem of balancing the risks and rewards in life to get the most out of every day remains the same. *Homo erectus* did it, you do it, and hopefully countless generations from now, humans will be doing their best at this as well.

Our journey is not over yet. We have far to go still, and many risks yet to face. We still have within us all the marvelous skills that got us this far, including our amazing adaptability, perhaps the most important skill we can have in a constantly changing world. We'll make mistakes along the way as individuals and probably as a species, but hopefully they'll be the kind we can learn from so we can get up the next day, once again, to take on this risky business called life.

NOTES

INTRODUCTION

1. Paul Slovic and Elke Weber, "Perception of Risk Posed by Extreme Events" (discussion, conference of the Risk Management Strategies in an Uncertain World, Palisades, NY, April 12–13, 2002).

CHAPTER 1. THE EVOLUTION OF RISK

1. Krishna Thompson, telephone interview and personal correspondence with the author, February 2012.

2. "Shark Attacks: Worry More about Lightning or Bees," *USNews.com*, August 31, 2010, http://www.usnews.com/science/articles/2010/08/31/shark-attacks-worry-more-about-lightning-or-bees (accessed August 16, 2012).

3. Jennifer Viegas, "Killer Sharks? Shark Perceptions Evolve from Cold-Blooded Killers to Environmental Saviors," Discovery, http://dsc.discovery.com/sharks/killer-sharks.html (accessed August 7, 2012).

4. S. E. Schlosser, "The King of the Sharks: A Native American Myth from Hawaii," AmericanFolklore.net, http://americanfolklore.net/folklore/2010/08/the_king_of_sharks.html (accessed August 7, 2012).

5. Ralph Collier, Shark Research Institute director and head of the Global Shark Attack File, telephone interviews with the author, December 2011 and January 2012.

6. Robert Sussman and Donna Hart, *Man the Hunted: Primate, Predators, and Human Evolution* (Cambridge, MA: Westview Press, 2005).

7. Robert Sussman, professor of anthropology at Washington University, St. Louis, Missouri, telephone interview and e-mail correspondence with the author, December 2011 and January 2012.

8. Sussman and Hart, *Man the Hunted*.

9. Gertrud Neaumann-Denzau and Helmut Denzau, "Examining Certain Aspects of Human-Tiger Conflict in the Sundarbans Forest, Bangladesh," *Tigerpaper* 37, no. 3

(July–September 2010): 1–11, http://www.fao.org/docrep/014/am243e/am243e00.pdf (accessed August 16, 2012).

10. Sussman and Hart, *Man the Hunted*.

11. Ibid.

12. Nicholas Nicastro, "Habitats for Humanity: Effects of Visual Affordance on the Evolution of Hominid Antipredator Communication," *Evolutionary Anthropology* 10 (2001): 153–57.

13. Gregory Juckett and John G. Hancox, "Venomous Snakebites in the United States: Management Review and Update," *American Family Physician* 65, no. 7 (April 1, 2002): 1367–75.

14. Geoffrey Brewer, "Snakes Top List of Americans' Fears," Gallup, March 19, 2001, http://www.gallup.com/poll/1891/snakes-top-list-americans-fears.aspx (accessed August 7, 2012).

15. Sussman and Hart, *Man the Hunted*.

16. M. Cook and S. Mineka, "Selective Associations in the Observational Conditioning of Fear in Rhesus Monkeys," *Journal of Experimental Psychology: Animal Behavior Processes* 16, no. 4 (October 1990): 372–89.

17. M. Cook and S. Mineka, "Selective Associations in the Origins of Phobic Fears and Their Implications for Behavior Therapy," in *Handbook of Behavior Therapy and Psychological Science: An Integrative Approach*, ed. P. Martin (Oxford: Pergamon Press, 1991), pp. 413–34.

18. Arne Ohman et al., "Emotion Drives Attention: Detecting the Snake in the Grass," *Journal of Experimental Psychology: General* 130, no. 3 (2001): 466–78.

19. Arne Ohman and Joaquim J. F. Soares, "Emotional Conditioning to Masked Stimuli: Expectancies for Aversive Outcomes following Nonrecognized Fear-Irrelevant Stimuli," *Journal of Experimental Psychology: General* 127, no. 1 (March 1998): 69–82.

20. Feinstein et al., "The Human Amygdala and the Induction and Experience of Fear," *Current Biology* 21, no. 1 (2011): 34–38.

21. Lynne Isbell, professor of anthropology and animal behavior at the University of California–Davis, telephone interview and e-mail correspondence with the author, May 2010–March 2012.

22. Lynne Isbell, *The Fruit, the Tree, and the Serpent: Why We See So Well* (Cambridge, MA: Harvard University Press, 2009).

23. Ulf Liszkowski, Malinda Carpenter, Tricia Striano, and Michael Tomasello, "Twelve- and 18-Month-Olds Point to Provide Information for Others," *Journal of Cognition and Development* 7, no. 2 (2006): 173–87.

24. Isbell, telephone interview with the author.

25. Thomas Headland and Harry Greene, "Hunter-Gatherers and Other Primates as Prey, Predators, and Competitors of Snakes," *Proceedings of the National Academy of Sciences* 108, no. 52 (December 27, 2011).

26. Lynne Isbell, "Why We See So Well," *Montreal Review*, October 2011.

27. Mark Changizi, *The Vision Revolution: How the Latest Research Overturns Everything We Thought We Knew about Human Vision* (Dallas, TX: BenBella Books, 2009).

28. Andy Cohen, joint professor of geosciences and ecology and evolutionary biology at the University of Arizona–Tucson, correspondence with the author, May 2010–November 2011.

29. "Climatic Fluctuations Drove Key Events in Human Evolution, Researchers Find," *Science Daily*, September 26, 2011.

CHAPTER 2. THE SEA OF DENIAL

1. Lynne Isbell, professor of anthropology and animal behavior at the University of California–Davis, telephone interview and e-mail correspondence with the author, May 2010–March 2012.

2. "Cuyahoga River Fire," *Ohio History Central*, July 1, 2005, http://www.ohiohistory central.org/entry.php?rec=1642 (accessed August 8, 2012).

3. For this and more NASA data, see http://climate.nasa.gov (accessed August 8, 2012).

4. Intergovernmental Panel on Climate Change (IPCC), www.ipcc.ch/.

5. For more on the impact of rising sea levels on coastal regions, see "Sea Level Rise: Ocean Levels Are Getting Higher—Can We Do Anything about It?" *National Geographic*, http://ocean.nationalgeographic.com/ocean/critical-issues-sea-level-rise/ (accessed August 8, 2012).

6. Dan Vergano, "2010 Was Record Year for Greenhouse Gas Emissions," *USA Today*, May 31, 2011.

7. Anthony Leiserowitz, "Communicating the Risks of Global Warming: American Risk Perceptions, Affective Images and Interpretive Communities," in *Creating a Climate for Change: Communicating Climate Change and Facilitating Social Change*, ed. Susanne Moser and Lisa Dilling (Cambridge, NY: Cambridge University Press, 2007).

8. Kari Marie Norgaard, *Living in Denial: Climate Change, Emotions, and Everyday Life* (Cambridge, MA: MIT Press, 2011).

9. Ibid.

10. Ibid.

11. Elke Weber, "Psychology: Climate Change Hits Home," *Nature Climate Change* 1 (2011): 25–26.

12. Alexa Spence et al., "Perceptions of Climate Change and Willingness to Save Energy related to Flood Experience," *Nature Climate Change* 1 (2011): 46–49.

13. Ibid.

14. J. B. Haufler, C. A. Mehl, and S. Yeats, *Climate Change: Anticipated Effects on Ecosystem Services and Potential Actions by the Alaska Region, U.S. Forest Service*, USDA report, 2010, http://www.fs.usda.gov/Internet/FSE_DOCUMENTS/fsbdev2_038171.pdf (accessed August 8, 2012).

15. Anthony Leiserowitz, "Does Experience Make a Difference? Alaskan Risk Perceptions of Climate Change" (PowerPoint presentation, Decision Center for a Desert City, Arizona State University, 2006).

16. Anthony Leiserowitz and Kenneth Broad, *Florida: Public Opinion on Climate Change*, 2008, http://environment.yale.edu/climate/files/Florida_Global_Warming_Opinion .pdf (accessed August 8, 2012).

17. Justin Gillis, "As Permafrost Thaws, Scientists Study the Risks," *New York Times*, December 16, 2011.

18. Nicholas Stern, "What Is the Economics of Climate Change?" *World Economics* 7, no. 2 (April–June 2006).

19. Bureau of Economic Analysis, "How Did the Recent GDP Revisions Change the Picture of the 2007–2009 Recession and the Recovery?" US Department of Commerce, August 5, 2011, http://www.bea.gov/faq/index.cfm?faq_id=1004 (accessed August 18, 2012).

20. American Psychological Association Task Force on the Interface between Psychology and Global Climate Change, *Psychology and Global Climate Change: Addressing a Multifaceted Phenomenon and Set of Challenges* (Washington, DC: American Psychological Association, March 2010), http://www.apa.org/science/about/publications/climate-change .aspx (accessed August 8, 2012).

21. Elke Weber, "Asymmetric Discounting in Intertemporal Choice: A Query Theory Account," *Psychological Science* 18 (2007): 516–23.

22. Ibid.

23. Dan Kahan, "Fixing the Communications Failure," *Nature* 463 (January 21, 2010): 296–97.

24. Kahan et al., "The Tragedy of the Risk-Perception Commons: Culture Conflict, Rationality Conflict, and Climate Change" (working paper no. 89, Cultural Cognition Project, 2011).

25. Eric Corey Freed, telephone interview with the author, February 2012.

26. Ibid.

27. Ibid.

28. "Leading Causes of Death," Centers for Disease Control and Prevention, 2012, http://www.cdc.gov/nchs/fastats/lcod.htm (accessed August 8, 2012).

29. Nick Parker, "15 Years on Nothing but Chicken Nuggets," *Sun*, January 26, 2012.

30. "Britons 'in Denial' over Heart Risk from Obesity and Smoking," *Daily Mail*, March 13, 2009, http://www.dailymail.co.uk/health/article-1161643/Britons-denial-heart -risk-obesity-smoking.html (accessed August 16, 2012).

31. S. Chapman, W. L. Wong, and W. Smith, "Self-Exempting Beliefs about Smoking

and Health: Differences between Smokers and Ex-smokers," *American Journal of Public Health* 83, no. 2 (February 1993): 215–19.

32. Robert Gramling et al., "Self-Rated Cardiovascular Risk and 15-Year Cardiovascular Mortality," *Annals of Family Medicine* 6, no. 4 (July 1, 2008): 302–306.

33. Jacob Levine et al., "The Role of Denial in Recovery from Coronary Heart Disease," *Psychosomatic Medicine* 49, no. 2 (March/April 1987): 109–17.

34. "The World's First Climate Change Refugees to Leave Island Due to Rising Sea Levels," *Daily Mail*, December 18, 2007.

CHAPTER 3. SUCKERS FOR A PRETTY FACE

1. Robert Reich, interviewed for the episode "Clinton," *American Experience* (PBS). For a transcript of the interview with Reich, see "Interview: Robert Reich," http://www.pbs .org/wgbh/americanexperience/features/interview/clinton-reich/ (accessed August 8, 2012).

2. "Sheryl" (pseud.; names in "Sheryl's" story changed to preserve anonymity), telephone interview with the author, February 9, 2012.

3. Ibid.

4. Ibid.

5. Arianne Cohen, author of *The Sex Diaries Project*, e-mail correspondence with the author, February 9, 2012.

6. Ibid.

7. Yolanda Stephen, personal correspondence with the author, December 2011.

8. Frank A. von Hippel, ed., *Tinbergen's Legacy in Behaviour: Sixty Years of Landmark Stickleback Papers* (Boston: Brill, 2010).

9. D. Shay, "Birds of Paradise: Family Paradisaeidae," UntamedScience.com, 2010, http://www.untamedscience.com/biodiversity/animals/chordates/birds/passerines/ paradisaeidae (accessed August 16, 2012).

10. Mark Ridley, *Evolution*, 3rd ed. (Malden, MA: Blackwell, 2009).

11. Jean-Guy J. Godin and Heather E. McDonough, "Predator Preference for Brightly Colored Males in the Guppy: A Viability Cost for a Sexually Selected Trait," *Behavioral Ecology* 14, no. 2 (2003): 194–200.

12. Ki Mae Heussner, "Science of Beauty: What Made Elizabeth Taylor So Attractive?" ABCNews.com, March 23, 2011, http://abcnews.go.com/Technology/elizabeth-taylor -science-great-beauty/story?id=13203775#.UCNO-U1lQ98 (accessed August 8, 2012).

13. G. Rhodes, "The Evolutionary Psychology of Facial Beauty," *Annual Review of Psychology* 57 (2006): 199–226.

14. Frank Marlowe, "The Nubility Hypothesis: The Human Breast as an Honest Signal of Residual Reproductive Value," *Human Nature* 9, no. 3 (1998): 263–71.

15. Deirdre Barret, *Supernormal Stimuli: How Primal Urges Overran Their Evolutionary Purpose* (New York: W. W. Norton, 2010).

16. G. Stulp et al., "A Curvilinear Effect of Height on Reproductive Success in Human Males," *Behavioral Ecology Sociobiology* 66 (2012): 375–84.

17. Randy Thornhill et al., "Male Symmetry and Female Orgasms, Human Female Orgasm and Mate Fluctuating Asymmetry," *Animal Behaviour* 50, no. 6 (1995): 1601–15.

18. Barrett, *Supernormal Stimuli.*

19. Ibid.

20. "Country Comparison: HIV/AIDS—Adult Prevalence Rate," CIA World Factbook, https://www.cia.gov/library/publications/the-world-factbook/rankorder/2155 rank.html (accessed August 16, 2012).

21. "HIV in the United States: At a Glance," Centers for Disease Control and Prevention, March 2012, http://www.cdc.gov/hiv/resources/factsheets/us.htm (accessed August 16, 2012).

22. I. Levy et al., "Men Who Have Sex with Men, Risk Behavior, and HIV Infection: Integrative Analysis of Clinical, Epidemiological, and Laboratory Databases," *Clinical Infectious Diseases* 52, no. 11 (June 2011): 1363.

23. H. Blanton and M. Gerrard, "Effect of Sexual Motivation on Men's Risk Perception for Sexually Transmitted Disease: There Must Be 50 Ways to Justify a Lover," *Health Psychology* 16, no. 4 (1997): 374–79.

24. "'Shady' Porn Site Practices Put Visitors at Risk," BBC News, June 11, 2010.

25. Liz Langley, *Crazy Little Thing: Why Love and Sex Drive Us Mad* (Berkeley, CA: Cleis Press, 2011).

26. Ibid., pp. 8–9.

27. Ibid., pp. 12–13.

28. Stephanie Rosenbloom, "A Simple Show of Hands," *New York Times*, October 5, 2006.

29. Daniel J. Kruger, "Economic Transition, Male Competition, and Sex Differences in Mortality Rates," *Evolutionary Psychology* 5, no. 2 (2007): 411–27.

30. Ibid.

31. "Life Expectancy at Birth," CIA World Factbook, https://www.cia.gov/library/publications/the-world-factbook/rankorder/2102rank.html (accessed August 9, 2012).

32. Kruger, "Economic Transition, Male Competition."

33. B. Pawlowski et al., "Sex Differences in Everyday Risk-Taking Behavior in Humans," *Evolutionary Psychology* 6, no. 1 (2008): 29–42.

34. Ibid.

35. Ibid.

36. Richard Ronay and William von Hippel, "The Presence of an Attractive Woman Elevates Testosterone and Physical Risk Taking in Young Men," *Social Psychological and Personality Science* 1, no. 1 (January 2010): 57–64.

37. John Coates and Joe Herbert, "Endogenous Steroids and Financial Risk Taking on a London Trading Floor," *Proceedings of the National Academy of Sciences of the United States of America* 105, no. 16 (April 22, 2008): 6167–72.

38. J. Archer, "Competition and Testosterone," *Neuroscience and Biobehavioral Reviews* 30, no. 3 (2006): 319–45.

CHAPTER 4. WHY WE BOTCH RARE RISKS

1. Dennis Mileti, professor emeritus at the University of Colorado–Boulder, telephone interview with the author, September 17, 2010.

2. Barack Obama, "Remarks by the President in a Discussion on Jobs and the Economy in Charlotte, North Carolina," Office of the Press Secretary, April 2, 2010, http://www.whitehouse.gov/the-press office/remarks-president-a-discussion-jobs-and-economy -charlotte-north-carolina (accessed September 9, 2012).

3. David Barstow, David Rohde, and Stephanie Saul, "Deepwater Horizon's Final Hours," *New York Times*, December 25, 2010, http://www.nytimes.com/2010/12/26/us/26spill.html?pagewanted=all (accessed August 16, 2012).

4. Mark Schrope, "The Lost Legacy of the Last Great Oil Spill," *Nature* 466 (2010): 304–305.

5. Robert Meyer, codirector of the University of Pennsylvania Wharton School, telephone interview with the author, September 2010.

6. Russell Hotten, "BP Attacked over 'Unsafe Culture,'" *Telegraph*, March 20, 2007.

7. US Department of Labor, "OSHA Issues Record-Breaking Fines to BP," news release, October 30, 2009, http://www.osha.gov/pls/oshaweb/owadisp.show_document?p _table=NEWS_RELEASES&p_id=16674 (accessed August 16, 2012).

8. Michael Isikoff and Michael Hirsh, "Slick Operator," *Newsweek Magazine, Daily Beast*, May 6, 2010, http://www.thedailybeast.com/newsweek/2010/05/07/slick-operator .html (accessed August 18, 2012).

9. Meyer, telephone interview with the author.

10. Barbara Kahn and Mary Luce, "Repeated-Adherence Protection Model: 'I'm OK, and It's a Hassle,'" *Journal of Public Policy and Marketing* 25, no. 1 (Spring 2006): 79–89.

11. David Perlman, "Huge State Quake Predicted within 30 Years," *San Francisco Chronicle*, April 15, 2008.

12. "Affordable Quake Insurance," *Los Angeles Times*, March 19, 2011, http://articles .latimes.com/2011/mar/19/opinion/la-ed-quakes-20110319 (accessed August 17, 2012).

13. Jason Fagone, "Masters of Disaster," *Wharton*, University of Pennsylvania, April 1, 2010, http://whartonmagazine.com/issues/spring-2010/masters-of-disaster/ (accessed August 18, 2012).

14. Meyer, telephone interview with the author.

15. Ibid.

16. Anthony Zachria, "Deaths Related to Hurricane Rita and Mass Evacuation," *Chest* 130, no. 4 (October 2006).

17. Meyer, telephone interview with the author.

18. Willie Drye, "Why Hurricane Ike's 'Certain Death' Warning Failed," *National Geographic*, September 26, 2008.

19. Meyer, telephone interview with the author.

20. Art Markman, professor in the Psychology Department at the University of Texas–Austin, telephone and e-mail interviews with the author, September 2010.

21. Ibid.

22. "Direct Hit by Ivan in New Orleans Could Mean a Modern Atlantis," *USA Today*, September 14, 2004.

23. Ralph Hertwig et al., "Decisions from Experience and the Effect of Rare Events in Risky Choice," *Psychological Science* 15, no. 8 (2004): 534–39.

24. Tim Folger, "Tsunami Science," *NationalGeographic.com*, February 2012, http://ngm.nationalgeographic.com/2012/02/tsunami/folger-text (accessed August 17, 2012).

25. Ibid.

26. Nassim Nicholas Taleb, *The Black Swan* (New York: Random House, 2007).

27. Ibid., pp. xxi–xxii.

28. Mileti, telephone interview with the author.

29. Ibid.

30. Ibid.

31. Ibid.

32. Ibid.

33. Frank Newport, "Nine Years after 9/11, Few See Terrorism as Top U.S. Problem," Gallup, September 10, 2010, http://www.gallup.com/poll/142961/nine-years-few-terrorism-top-problem.aspx (accessed August 17, 2012). Results include summary of past terrorism-poll results.

34. Ibid.

35. L. C. Johnson, "The Declining Terrorist Threat," *New York Times*, July 10, 2001.

36. Michael Rothschild, "Terrorism and You—the Real Odds," *Washington Post*, November 25, 2001.

37. Paul Campos, "Undressing the Terrorism Threat," *Wall Street Journal*, January 9, 2010.

38. Leonie Huddy et al., "The Consequences of Terrorism: Disentangling the Effects of Personal and National Threat," *Political Psychology* 23, no. 3 (2002); Leonie Huddy, Nadia Khatib, and Theresa Capelos, "The Polls—Trends," *Public Opinion Quarterly* 66 (2002): 418–50, http://ms.cc.sunysb.edu/~lhuddy/HuddyKhatibCapelosPOQ.pdf (accessed August 18, 2012).

39. Joseph Carroll, "Americans' Terrorism Worries Five Years after 9/11," Gallup, September 11, 2006, http://www.gallup.com/poll/24412/americans-terrorism-worries-five-years-after-911.aspx (accessed August 17, 2012).

40. Woods et al., "Terrorism Risk Perceptions and Proximity to Primary Terrorist Targets: How Close Is Too Close?" *Human Ecology Review* 15, no. 1 (2008): 63.

41. For a transcript of the video *7 Signs of Terrorism*, see http://www.emich.edu/cerns/downloads/cert/Seven_Signs_of_Terrorism.htm (accessed August 18, 2012).

42. Timur Kuran and Cass R. Sunstein, "Availability Cascades and Risk Regulation," *Stanford Law Review* 51, no. 4 (April 1999): 683–768.

43. "Defenses Upgraded, but Some Seek More," *Wall Street Journal*, August 28, 2010.

44. Newport, "Nine Years after 9/11."

45. Meyer, telephone interview with the author.

46. Ibid.

47. Ibid.

CHAPTER 5. THE BALANCING ACT

1. Michael Cooper, "Happy Motoring: Traffic Deaths at 61-Year Low," *New York Times*, April 1, 2011, http://www.nytimes.com/2011/04/01/us/01driving.html (accessed August 20, 2012).

2. World Health Organization, "Global Status Report on Road Safety," http://www.who.int/violence_injury_prevention/road_safety_status/en/ (accessed August 19, 2012).

3. Centers for Disease Control, "Teen Drivers: Fact Sheet," 2010, http://www.cdc.gov/motorvehiclesafety/teen_drivers/teendrivers_factsheet.html (accessed August 19, 2012).

4. Daniel Lapsley and P. Hill, "Subjective Invulnerability, Optimism Bias and Adjustment in Emerging Adulthood," *Journal of Youth and Adolescence* 39, no. 8 (August 2010): 847–57.

5. Tara Parker-Pope, "Teenagers, Friends, and Bad Decisions," *New York Times*, February 3, 2011, http://well.blogs.nytimes.com/2011/02/03/teenagers-friends-and-bad-decisions/ (accessed August 19, 2012).

6. Laurence Steinberg, "Risk Taking in Adolescence: New Perspectives from Brain and Behavioral Science," *Current Directions in Psychological Science* 16, no. 2 (April 2007).

7. J. Chein et al., "Peers Increase Adolescent Risk Taking by Enhancing Activity in the Brain's Reward Circuitry," *Developmental Science* 14, no. 2 (March 2011): F1–F10.

8. Giovanni Laviola et al., "Risk-Taking Behavior in Adolescent Mice: Psychobiological Determinants and Early Epigenetic Influence," *Neuroscience and Biobehavioral Review* 27, nos. 1–2 (January–March 2003): 19–31.

9. Steinberg, "Risk Taking in Adolescence."

10. Elizabeth Landau, "Why Teens Are Wired for Risk," *CNN*, October 19, 2011, http://www.cnn.com/2011/10/19/health/mental-health/teen-brain-impulses/index.html (accessed August 19, 2012).

11. Lapsley and Hill, "Subjective Invulnerability."

12. Sheryl Ubelacker, "Teens Misjudge Risk to Selves from Accidents: Study," *Toronto Star*, August 14, 2008.

13. Ibid.

14. Ibid.

15. Gerald Wilde, *Target Risk: Dealing with the Danger of Death, Disease and Damage in Everyday Decisions* (Toronto: PDE Publications, 1994), available online at psyc.queensu.ca/target/index.html; the phrase *risk thermostat* is mentioned in Wilde's book.

16. Suzanne Collins, *The Hunger Games* (New York: Scholastic, 2009).

17. Wilde, *Target Risk*.

18. Ibid.

19. Thomas Wenzel and Marc Ross, "The Effects of Vehicle Model and Driver Behavior on Risk," *Accident Analysis and Prevention* 37 (2005): 479–94.

20. Ibid.

21. J. A. Thomas and D. Walton, "Measuring Perceived Risk: Self-Reported and Actual Hand Positions of SUV and Car Drivers," *Transportation Research Part F: Traffic Psychology and Behaviour* 10, no. 3 (May 2007): 201–207.

22. Lesley Walker, Jonathan Williams, and Konrad Jamrozik, "Unsafe Driving Behaviour and Four Wheel Drive Vehicles: Observational Study," *BMJ* 333 (July 6, 2006): 71.

23. Wilde, *Target Risk*. See specifically the discussion in chapter 7.

24. Ibid.

25. R. M. Harano and D. E. Hubert, *An Evaluation of California's 'Good Driver' Incentive Program: Report No. 6* (Sacramento: California Division of Highways, 1974).

26. Ibid.

27. James Angel, associate professor of finance at the McDonough School of Business, Georgetown University, personal correspondence with the author, March 8, 2012.

28. Ibid.

29. Abbigail J. Chiodo and Michael T. Owyang, "Low Unemployment: Old Dogs or New Tricks?" *Regional Economist* (October 2001). Read the article at http://www.stlouisfed.org/publications/re/articles/?id=455.

30. Anthony Accetta, founder of the Accetta Group, telephone and e-mail interviews with the author, March 2012.

31. Ibid.

32. Ibid., italics added.

33. Ibid.

34. "Buffett Warns on Investment 'Time Bomb,'" *BBC News*, March 4, 2003.

35. Angel, personal correspondence with the author.

36. Madison Park, "Half of Americans Use Supplements," *CNN*, April 13, 2011, http://www.cnn.com/2011/HEALTH/04/13/supplements.dietary/index.html (accessed August 22, 2012).

37. Wen-Bin Chiou et al., "Ironic Effects of Dietary Supplementation: Illusory Invulnerability Created by Taking Dietary Supplements Licenses Health-Risk Behaviors," *Psychological Science* 22, no. 8 (2011): 1081–86.

38. "Waistlines in People, Glucose Levels in Mice Hint at Sweeteners' Effects: Related Studies Point to the Illusion of the Artificial," University of Texas Health Science Center–San Antonio, June 27, 2011, http://www.uthscsa.edu/hscnews/single format2.asp?newID=3861 (accessed August 19, 2012).

39. Emma Jane Kirby, "Should All Skiers Wear Helmets?" *BBC News*, January 26, 2012, http://www.bbc.co.uk/news/health-16715883 (accessed August 22, 2012).

40. Lisa A. Eaton and Seth C. Kalichman "Risk Compensation in HIV Prevention: Implications for Vaccines, Microbicides, and Other Biomedical HIV Prevention Technologies," *Current HIV/AIDS Reports* 4, no. 4 (December 2007): 165–72.

41. T. L. Mullins et al., "Adolescent Perceptions of Risk and Need for Safer Sexual Behaviors after First Human Papillomavirus Vaccination," *Archives of Pediatric & Adolescent Medicine* 166, no. 1 (2012): 82–88.

42. "Many Teen Girls Mistakenly Think HPV Vaccines Cut Risk for All STDs," HealthyDay.com, January 5, 2012.

43. "Human Error Is the Biggest Obstacle to 100 Percent Flight Safety," *Denver Post*, February 14, 2010.

44. Robert Sumwalt, *Current Issues with Air Medical Transportation: EMS Helicopter Safety* (presentation to the National Transportation Safety Board, May 4, 2011), http://www.ntsb.gov/doclib/speeches/sumwalt/sumwalt_050411.pdf (accessed August 20, 2012).

45. "Medical-Helicopter Study on Safety Splits Industry," *Wall Street Journal*, April 20, 2009.

46. Steve Greene, telephone and e-mail interviews with the author, March 2012.

47. "IT May Never Make Safeguard of Sensitive Information Foolproof," *Economic Times*, February 10, 2008.

CHAPTER 6. LOSING CONTROL AND GAINING FEAR

1. Elliot D. Cohen, "The Fear of Losing Control," *Psychology Today*, May 22, 2011.

2. Paul Slovic, professor of psychology at the University of Oregon and president of Decision Research Group, telephone interview with the author, December 2011.

3. Ibid.

4. Martin Seligman, "Learned Helplessness," *Annual Review of Medicine* 23 (February 1972): 407–12.

5. "Skydiving Safety," United States Parachute Association, 2011, http://www.uspa .org/AboutSkydiving/SkydivingSafety/tabid/526/Default.aspx (accessed August 27, 2012).

6. Chauncey Starr, "Social Benefits versus Technological Risks," *Science* 165, no. 3899 (September 19, 1969): 1232–38.

7. "With Rise in Radiation Exposure, Experts Urge Caution on Tests," *New York Times*, June 19, 2007.

8. "Japan Radiation Risk to California Is Downplayed," *Los Angeles Times*, March 16, 2011.

9. "Chernobyl Nuclear Accident: Figures for Deaths and Cancers Still in Dispute," *Guardian*, January 10, 2010.

10. R. E. Adams et al., "Stress and Well-Being in Mothers of Young Children 11 Years after the Chornobyl Nuclear Power Plant Accident," *Psychological Medicine* 32 (2002): 143–56.

11. H. M. Ginzburg, "The Psychological Consequences of the Chernobyl Accident— Findings from the International Atomic Energy Agency Study," *Public Health Report* 108, no. 2 (March–April 1993): 184–92.

12. "JetBlue Flight Diverted after Captain's 'Erratic' Behavior," *CNN*, March 27, 2012.

13. Ibid.

14. David Hunter et al., *Locus of Control, Risk Orientation, and Decision Making among U.S. Army Aviators* (Arlington, VA: US Army Research Institute for the Behavioral and Social Sciences, October 2009).

15. Guohua Li and Susan P. Baker, "Crash Risk in General Aviation," *Journal of the American Medical Association* 297, no. 14 (April 11, 2007): 1596–98.

16. Ibid.

17. David Hunter, "Risk Perception among General Aviation Pilots," *International Journal of Aviation Psychology* 16, no. 2 (2006): 135–44.

18. Keryn A. Pauley et al., "Implicit Perceptions of Risk and Anxiety and Pilot Involvement in Hazardous Events, Human Factors," *Human Factors* 50, no. 5 (October 2008): 723–33.

19. Frank McKenna, "It Won't Happen to Me: Unrealistic Optimism or Illusion of Control?" *British Journal of Psychology* 84, no. 1 (1993): 39–50.

20. Ibid.

21. Ibid.

22. Dan Nainan, e-mail interview with the author, March 2012.

23. Edie Raether, e-mail interview with the author, March 2012.

24. Ibid.

25. Kevin Foster and Hanna Kokko, "The Evolution of Superstitious and Superstition-

like Behaviour," *Proceedings of the Royal Society B: Biological Sciences* 276, no. 1654 (2009): 31–37.

26. Ibid.

27. Ibid.

28. Robert Ladouceur, "Perceptions among Pathological and Nonpathological Gamblers," *Addictive Behaviors* 29 (2004): 555–65.

29. Jennie Finch, e-mail correspondence with the author, March 2012.

30. Ibid.

31. Ibid.

32. Lysann Damisch, "Keep Your Fingers Crossed! The Influence of Superstition on Subsequent Task Performance and Its Mediating Mechanism" (doctoral dissertation, University of Cologne, Germany, 2008).

33. Meghan Collins, "Traders Ward Off Evil Spirits," *CNN*, October 31, 2003.

34. Carl Bialik, "The Power of Lucky Charms," *Wall Street Journal*, April 28, 2010.

35. James T. Reddy, "Chinese Investors Crunching Numbers Are Glad to See 8s," *Wall Street Journal*, May 24, 2007.

36. Adam Galinsky and Jennifer Whitson, "Lacking Control Increases Illusory Pattern Perception," *Science* 322 (October 3, 2008): 115–17.

37. Ibid.

38. Ibid.

39. Ibid.

40. Ibid.

CHAPTER 7. THE STORY OF VACCINES AND AUTISM

1. Susan Dominus, "The Crash and Burn of an Autism Guru," *New York Times*, April 20, 2011.

2. Wakefield et al., "RETRACTED: Ileal-Lymphoid-Nodular Hyperplasia, Non-specific Colitis, and Pervasive Developmental Disorder in Children," *Lancet* 351, no. 9103 (February 28, 1998): 637–41.

3. Denis Campbell, "I Told the Truth All Along, Says Doctor at Heart of Autism Row," *Observer*, July 7, 2007, http://www.guardian.co.uk/society/2007/jul/08/health.medicineandhealth1 (accessed August 24, 2012).

4. "Unvaccinated Kids behind Largest U.S. Measles Outbreak in Years: Study," *U.S. News and World Report*, October 20, 2011.

5. Robin Dunbar, professor of evolutionary psychology at Oxford University, personal correspondence with the author, January 2012.

6. Gerd Gigerenzer, "Helping Physicians Understand Screening Tests Will Improve Health Care," *Association for Psychological Science* 20, nos. 37–38 (November 2007).

7. Markus Feufel, e-mail correspondence with the author, March 2012.

8. See the fact box on mammography from the Harding Center website at http://www.harding-center.com/fact-boxes/mammography (accessed August 20, 2012).

9. See the fact box on prostate cancer screening from the Harding Center website at http://www.harding-center.com/fact-boxes/psa-screening (accessed August 20, 2012).

10. Matt Kalla, live and e-mail interviews with the author, San Diego, CA, February 2012.

11. Ibid.

12. Yuval Rottenstreich and C. K. Hsee, "Money, Kisses, and Electric Shocks," *Psychological Science* 12, no. 3 (May 2001): 185–90.

13. Yuval Rottenstreich, professor of management and organizations at the NYU Stern School of Business, personal correspondence with the author, March 2012.

14. Ibid.

15. Cass R. Sunstein and Richard Zeckhauser, "Dreadful Possibilities, Neglected Probabilities," in *The Irrational Economist: Making Decisions in a Dangerous World*, ed. Erwann Michel-Kerjan and Paul Slovic (New York: Public Affairs Books, 2010).

16. Daniel Kahneman, Jack Knetsch, and Richard Thaler, "Experimental Tests of the Endowment Effect and the Coase Theorem," *Journal of Political Economy* 98, no. 6 (December 1990): 1325–48.

17. Daniel Kahneman and Amos Tversky, "Prospect Theory: An Analysis of Decision under Risk," *Econometrica* 47, no. 2 (March 1979): 263–92.

18. Paul Slovic, "The More Who Die, the Less We Care," in Michel-Kerjan and Slovic, *Irrational Economist*, pp. 30–40.

19. Ibid.

20. Deborah A. Small, George Loewenstein, and Paul Slovic, "Sympathy and Callousness: The Impact of Deliberative Thought on Donations to Identifiable and Statistical Victims," *Organizational Behavior and Human Decision Processes* 102 (March 2007): 143–53.

21. Paul Slovic, professor of psychology at the University of Oregon and president of Decision Research Group, telephone interview with the author, December 2011.

22. Ibid.

CHAPTER 8. THE THING WE FEAR MORE THAN DEATH

1. Roy Baumeister and Mark Leery, "The Need to Belong: Desire for Interpersonal Attachments as a Fundamental Human Motivation," *Psychological Bulletin* 117, no. 3 (May 1995): 497–529.

2. Prof. Robin Dunbar, director of the Institute of Cognitive and Evolutionary

Anthropology at the University of Oxford, personal correspondence with the author, January 2012.

3. Kipling "Kip" Williams, professor of psychological sciences at Purdue University, personal correspondence with the author, January–March 2012.

4. Ibid.

5. Kipling Williams, "Ostracism: The Kiss of Social Death," *Social and Personality Psychology Compass* 1, no. 1 (2007): 236–47.

6. Kipling Williams, "The Pain of Exclusion," *Scientific American Mind* 21, no. 6 (January/February 2011): 32–39.

7. Ibid., p. 35.

8. Williams, "Ostracism."

9. Henry David Thoreau, *Walden* (Boston: Houghton Mifflin, 1854).

10. Alston Chase, "Harvard and the Making of the Unibomber," *Atlantic Online*, June 2000, http://www.theatlantic.com/past/docs/issues/2000/06/chase.htm (accessed August 24, 2012).

11. Theodore Kaczynski, *The Unabomber Manifesto: Industrial Society and Its Future* (Minneapolis, MN: Filiquarian, 1995).

12. Ryan Devereaux, "Solitary Confinement on Trial: Senators Hear from Experts on Prison Reform," *Guardian*, June 19, 2012, http://www.guardian.co.uk/world/2012/jun/19/solitary-confinement-trial-us-senators (accessed September 12, 2012).

13. Craig Haney, professor of psychology, University of California–Santa Cruz, quoted in Brandon Keim, "Solitary Confinement: The Invisible Torture," *Wired*, April 29, 2009, http://www.wired.com/wiredscience/2009/04/solitaryconfinement/ (accessed August 20, 2012).

14. Ibid.

15. Williams, "Ostracism"; C. N. DeWall et al., "Acetaminophen Reduces Social Pain: Behavioral and Neural Evidence," *Psychological Science* 21, no. 7 (July 2010): 931–37.

16. Naomi Eisenberger and Matthew D. Lieberman, "Why Rejection Hurts: A Common Neural Alarm System for Physical and Social Pain," *Trends in Cognitive Sciences* 8, no. 7 (July 2004).

17. Naomi Eisenberger, quoted in Richard Alleyne, "Breaking Up Hurts Just Like Physical Pain," *Telegraph*, December 4, 2009.

18. Anne Ryman, "UA Study: Divorce Can Raise Risk of Early Death," *Arizona Republic*, *USAToday.com*, January 10, 2012, http://www.usatoday.com/news/health/wellness/story/2012-01-10/UA-study-Divorce-can-raise-risk-of-early-death/52478544/1 (accessed August 20, 2012).

19. Robin Dunbar, "You've Got to Have (150) Friends," *New York Times*, December 25, 2010.

20. Robin Dunbar, Louise Barrett, and John Lycett, *Evolutionary Psychology: A Beginner's Guide—Human Behaviour, Evolution, and the Mind* (Oxford, UK: OneWorld, 2007).

21. Ibid.

22. "Gossip," *This American Life*, NPR, August 26, 2011. For a transcript of the episode, see http://www.thisamericanlife.org/radio-archives/episode/444/transcript (accessed August 20, 2012).

23. "Paul Zak: Trust, Morality—and Oxytocin," TEDTalks video, 16:34, lecture by neuroeconomist Paul Zak filmed July 2011, posted by "TEDGlobal 2011," November 2011, http://www.ted.com/talks/paul_zak_trust_morality_and_oxytocin.html (accessed August 20, 2012).

24. Ibid.

25. Ibid.

26. Ibid.

27. Ibid.

28. Ibid.

29. "Facebook 'Linked to Rise in Syphilis,'" *Telegraph*, March 24, 2010.

30. Alina Adams, personal correspondence with the author, March 2012.

31. Ibid.

32. Alina Adams, *Soap Opera 451: A Time Capsule of Daytime Drama's Greatest Moments* (n.p.: Alina Adams Media, 2011).

33. Todd Kashdan and Jon Elhai, "Social Anxiety and Positive Outcome Expectancies on Risk-Taking Behaviors," *Cognitive Therapy and Research* 30 (2006): 749–61.

34. Todd Kashdan and P. E. McKnight, "The Darker Side of Social Anxiety: When Aggressive Impulsivity Prevails over Shy Inhibition," *Current Directions in Psychological Science* 19, no. 1 (2010): 47–50.

35. Gregg Steinberg, telephone interview with the author, January 2012.

36. Marjorie Asturias, Dallas, Texas, personal correspondence with the author, March 2012.

37. Signe A. Dayhoff, e-mail correspondence with the author, March 2012.

38. Ibid.

39. Thomas H. Maugh II, "July Is the Worst Month to Check into a Teaching Hospital, UCSD Researchers Say," *Los Angeles Times*, June 2, 2010.

40. Thomas Blass, "The Man Who Shocked the World," *Psychology Today*, March 1, 2002.

41. Stanley Milgram, "The Perils of Obedience," *Harper's Magazine*, December 1973.

42. John Schwartz, "Simulated Prison in '71 Showed a Fine Line between 'Normal' and 'Monster,'" *New York Times*, May 6, 2004.

43. Richard Johnston, "The Battle against White-Collar Crime," *USA Today*, January 1, 2002.

44. Bill Ibelle, "The Whistleblower's Lawyer," from February 4, 2009, testimony to Congress, in "Special Section: Stories of Accountability," supplement, *Harvard Law*

Bulletin (2009), http://www.law.harvard.edu/news/bulletin/2009/summer/feature_1-side3 .php (accessed August 20, 2012).

45. Jim Angleton, e-mail interview with the author, March 2012.

CHAPTER 9. WHAT DO CAVE DIVING, NAKED SKYDIVING, AND ENTREPRENEURSHIP HAVE IN COMMON?

1. Jill Heinerth, personal correspondence with the author, October 2011–March 2012.

2. Ibid.

3. Ibid.

4. Ibid.

5. D. Comings and K. Blum, "Reward Deficiency Syndrome: Genetic Aspects of Behavioral Disorders," *Progress in Brain Research* 126 (2000): 325–41.

6. Brad Wible, "For Some Parkinson's Drugs, the Side Effect May Be Gambling," *Los Angeles Times*, July 12, 2005.

7. Dave Finlay, "Drug Trial Guinea Pig Loses Cash Bid," *Daily Record*, May 25, 2011.

8. K. J. Klos et al., "Pathological Hypersexuality Predominantly Linked to Adjuvant Dopamine Agonist Therapy in Parkinson's Disease and Multiple System Atrophy," *Parkinsonism & Related Disorders* 11, no. 6 (September 2005): 381–86.

9. "Risk of Dying and Sporting Activities," *Bandolier*, http://www.medicine.ox.ac .uk/bandolier/booth/risk/sports.html (accessed August 21, 2012).

10. Roberta Mancino, telephone interview with the author, January 2012.

11. Ibid.

12. J. E. Joseph et al., "Neural Correlates of Emotional Reactivity in Sensation Seeking," *Psychological Science* 20, no. 2 (February 2009): 215–23.

13. J. Reuter et al., "Pathological Gambling Is Linked to Reduced Activation of the Mesolimbic Reward System," *Nature Neuroscience* 8, no. 2 (February 2005): 147–48.

14. A. Lawrence et al., "The Innovative Brain," *Nature* 456, no. 7219 (November 13, 2008).

15. David Linden, *The Compass of Pleasure: How Our Brains Make Fatty Foods, Orgasm, Exercise, Marijuana, Generosity, Vodka, Learning, and Gambling Feel So Good* (New York: Viking, 2011).

CHAPTER 10. LIVING IN A RISKY WORLD

1. Aaron Klein, CEO of Riskalyze, telephone interview with the author, January 2012.

2. Jennifer Steil, e-mail correspondence with the author, March 2012.

3. Ibid.

4. Paul Slovic, professor of psychology at the University of Oregon and president of Decision Research Group, telephone interview with the author, December 2011.

INDEX